A Complete Study Guide For Technician, General, Extra Class Ham Radio Exams, and the Volunteer Examiner's Test:

Including the Correct Answers to All Questions, With Chapters on Basic Theory, Rules, and Regulations

by Joseph Lumpkin

AB4AN

The Complete Study Guide For All Amateur Radio Tests

A Complete Study Guide For Technician, General, Extra Class Ham Radio Exams, and the Volunteer Examiner's Test: Including the Correct Answers to All Questions, With Chapters on Basic Theory, Rules, and Regulations

Copyright © 2012 Joseph Lumpkin

All rights reserved.

Printed in the United States of America. No part of this book may be used or reproduced in any manner whatsoever without written permission except in the case of brief quotations embodied in critical articles and reviews.

Fifth Estate Publishers,

Post Office Box 116, Blountsville, AL 35031.

First Printing, 2012

Cover Design by An Quigley

Printed on acid-free paper

Library of Congress Control No: 2012951715

ISBN: 9781936533305

Fifth Estate, 2012

The Complete Study Guide For All Amateur Radio Tests

Table of Contents

Preface	5
Introduction	6
Technician Class	8
General Class	72
Extra Class	144
Volunteer Examiner	255
General Hints for Tests	262
Theory and Formulas	264
FCC Part 97	386

The Complete Study Guide For All Amateur Radio Tests

Preface

Don't think of a black cat – Don't think of a black cat – Do not think of a black cat… The brain does not process negatives very well. It takes in information first and then attempts to filter the results through logic later. To expose someone to the incorrect answers on tests is to place fraudulent information in memory and then say to the brain, "Don't remember these answers." It is for this reason the only answers you will encounter in this book are the correct ones.

Even though there are many answers to remember, you only have to remember answers for one test at a time. Out of all the answers to each test there are many questions that are similar with answers that are the same. There are also patterns within the answers since many questions are related. Jot down these patterns in the margins as you see them. It is not difficult if you approach the process one step at a time. Look for the same answers being repeated for various questions. Study the questions and answers in the front of the book. When it comes time to take the test the answers will leap off the page at you. You will recognize them at once.

The more difficult concepts are explained in further detail in the back of the book. Not all subjects are covered in the explanation section. Some subjects, such as common courtesy and malicious interference are a matter of common sense. The more math-intensive subjects are given attention in deference to those who have the study time and wish know more than just the correct answers. For most of us, living a life that is too busy for our own good, the first section of the book will be enough to pass the tests and move forward. Study the answers. There will become familiar and will be very recognizable when you take the tests.

Presenting the Questions and Answers to the Technician, General, and Extra Class Ham Radio Exams, as well as the Volunteer Examiner's Test

Introduction

Amateur Radio, or Ham radio, as most know it, is a thriving community. Over the years of growth in technology Ham Radio has embraced progress and contributed in ways that stretch the imagination. Hams have been involved in technology and events that changed history. From the ability to transmit and receive the first images from the moon to today's satellite communications, Ham Radio operators have contributed. Our numbers continue to grow. There are now over 700,000 licensed radio amateurs in the U.S. They embrace the hobby in dozens of ways. From social communications to the science of antennas, the various digital modes and satellite communications, Ham Radio has something to offer everyone. If you have been putting off entry into the world of amateur radio because of the tests, now is the time to step up and go for it.

There are no formulas to memorize, no math, no electronics or theories to learn. This manual contains the answers to all possible questions in all tests from Technician Class to Extra Class license tests. In addition, the manual also provides the answers to the Volunteer Examiner's test. However, many people need to know "why" and many need a framework on which the answers are related. That is why this book is arranged so that theory, formulas, rules, and regulations are explained in separate chapters.

Mores Code is no longer required. All you have to do is pass a series of multiple-choice exams. Simply study this manual and take the tests. The answers will be easily recognizable in the multiple-choice tests.

Each of the tests has a "pool" of pre-defined questions that are selected at random to make up the actual test. The questions are presented in a random order and the answers are also randomized from "A" to "D" in the multiple-choice style. You can't memorize the fact that question T1A04 is answer C, but you can memorize the actual answer. Your mind will recognize the correct answer. Pay no attention to the LETTER of the correct answer. Simply memorize the question and answer.

If you are not sure you are ready, there are several sites that provide practice tests to allow you to become comfortable and confident with the tests.

http://www.eham.net/exams/

http://www.eham.net/exams/

http://www.w8mhb.com/exam/

http://www.w8mhb.com/exam/

I used this method of memorization and practice testing to achieve a score of 100% on my Extra class test. You can too.

After you have prepared for your tests you can contact a Ham in your area for test dates or simply go to www.arrl.org for information on all license tests, dates, and addresses. After you pass the license exams you can go on line and advance to the final step, which I hope you will take. It is the Volunteer Examiner's Test. The answers to the VE exam are also in this manual.

Contained in the back of the book are general theories and concepts of electronics and antenna design. The formulas and diagrams are found in various books and websites that teach electronic, physics, and antenna design and are selected and brought together so the reader will have all the information needed to pass the licensing tests in one place. By addressing the major concepts a general understanding can be acquired so the reader can have basic concepts on which to hang and understand the answers to the test.

At the end of this book you will also find a chapter with hints to help you take and pass your tests. Finally, there are chapters for those who which to understand more of the math, electronics and physics behind the answers. To derive the correct answers to the test questions one can remember the answers or apply the formulas. This book attempts to accommodate both approaches. Lastly, FCC part 97, which applies to Amateur Radio, is included in the back of this book. With the information and answers in the book you will be able to pass all tests required by the FCC.

Good Luck

73's

Joseph Lumpkin

AB4AN

The Complete Study Guide For All Amateur Radio Tests

TECHNICIAN CLASS

Second Release - January 29, 2010

Element 2 Technician Class Question Pool

To be Effective July 1, 2010 for Examinations

(With Errata to first release version dated January 4, 2010)

The Technician class (Element 2) Pool is effective July 1, 2010 and is valid until June 30, 2014. The question pool contains seven schematic diagrams.

After the initial release of this question pool January 4, 2010, minor changes were made in the wording of questions and answers to correct typos and/or clarify meaning. The changes incorporated in the text of the pool are reflected in the questions below.

SUBELEMENT T1 – FCC Rules, descriptions and definitions for the amateur radio service, operator and station license responsibilities - [6 Exam Questions - 6 Groups]

T1A - Amateur Radio services; purpose of the amateur service, amateur-satellite service, operator/primary station license grant, where FCC rules are codified, basis and purpose of FCC rules, meanings of basic terms used in FCC rules

T1A01 (D) [97.3(a)(4)]

For whom is the Amateur Radio Service intended?

D. Persons who are interested in radio technique solely with a personal aim and without pecuniary interest

T1A02 (C) [97.1]

What agency regulates and enforces the rules for the Amateur Radio Service in the United States?

C. The FCC

T1A03 (D)

Which part of the FCC rules contains the rules and regulations governing the Amateur Radio Service?

D. Part 97

T1A04 (C) [97.3(a)(23)]

Which of the following meets the FCC definition of harmful interference?

C. That which seriously degrades, obstructs, or repeatedly interrupts a radio communication service operating in accordance with the Radio Regulations

T1A05 (D) [97.3(a)(40)]

What is the FCC Part 97 definition of a space station?

D. An amateur station located more than 50 km above the Earth's surface

T1A06 (C) [97.3(a)(43)]

What is the FCC Part 97 definition of telecommand?

C. A one-way transmission to initiate, modify or terminate functions of a device at a distance

T1A07 (C) [97.3(a)(45)]

What is the FCC Part 97 definition of telemetry?

C. A one-way transmission of measurements at a distance from the measuring instrument

T1A08 (B) [97.3(a)(22)]

Which of the following entities recommends transmit/receive channels and other parameters for auxiliary and repeater stations?

B. Frequency Coordinator

T1A09 (C) [97.3(a)(22)]

Who selects a Frequency Coordinator?

C. Amateur operators in a local or regional area whose stations are eligible to be auxiliary or repeater stations

T1A10 (A) [97.3(a)(5)]

What is the FCC Part 97 definition of an amateur station?

A. A station in an Amateur Radio Service consisting of the apparatus necessary for carrying on radio communications

T1A11 (C) [97.3(a)(7)]

Which of the following stations transmits signals over the air from a remote receive site to a repeater for retransmission?

C. Auxiliary station

T1B - Authorized frequencies; frequency allocations, ITU regions, emission type, restricted sub-bands, spectrum sharing, transmissions near band edges

T1B01 (B) [97.3(a)(28)]

What is the ITU?

B. A United Nations agency for information and communication technology issues

T1B02 (B)

North American amateur stations are located in which ITU region?

B. Region 2

T1B03 (B) [97.301(a)]

Which frequency is within the 6 meter band?

B. 52.525 MHz

T1B04 (A) [97.301(a)]

Which amateur band are you using when your station is transmitting on 146.52 MHz?

A. 2 meter band

T1B05 (C) [97.301(a)]

Which 70 cm frequency is authorized to a Technician Class license holder operating in ITU Region 2?

C. 443.350 MHz

T1B06 (B) [97.301(a)]

Which 23 cm frequency is authorized to a Technician Class operator license?

B. 1296 MHz

T1B07 (D) [97.301(a)]

What amateur band are you using if you are transmitting on 223.50 MHz?

D. 1.25 meter band

T1B08 (C) [97.303]

What do the FCC rules mean when an amateur frequency band is said to be available on a secondary basis?

C. Amateurs may not cause harmful interference to primary users

T1B09 (D) [97.101(a)]

Why should you not set your transmit frequency to be exactly at the edge of an amateur band or sub-band?

A. To allow for calibration error in the transmitter frequency display

B. So that modulation sidebands do not extend beyond the band edge

C. To allow for transmitter frequency drift

D. All of these choices are correct

To allow for calibration error in the transmitter frequency display

So that modulation sidebands do not extend beyond the band edge

To allow for transmitter frequency drift

T1B10 (C) [97.305(c)]

Which of the bands available to Technician Class operators have mode-restricted sub-bands?

C. The 6 meter, 2 meter, and 1.25 meter bands

T1B11 (A) [97.305 (a)(c)]

What emission modes are permitted in the mode-restricted sub-bands at 50.0 to 50.1 MHz and 144.0 to 144.1 MHz?

A. CW only

T1C - Operator classes and station call signs; operator classes, sequential, special event, and vanity call sign systems, international communications, reciprocal operation, station license and licensee, places where the amateur service is regulated by the FCC, name and address on ULS, license term, renewal, grace period

T1C01 (C) [97.3(a)(11)(iii)]

Which type of call sign has a single letter in both the prefix and suffix?

C. Special event

T1C02 (B)

Which of the following is a valid US amateur radio station call sign?

B. W3ABC

T1C03 (A) [97.117]

What types of international communications are permitted by an FCC-licensed amateur station?

A. Communications incidental to the purposes of the amateur service and remarks of a personal character

T1C04 (A)

When are you allowed to operate your amateur station in a foreign country?

A. When the foreign country authorizes it

T1C05 (A) [97.303(h)]

What must you do if you are operating on the 23 cm band and learn that you are interfering with a radiolocation station outside the United States?

A. Stop operating or take steps to eliminate the harmful interference

T1C06 (D) [97.5(a)(2)]

From which of the following may an FCC-licensed amateur station transmit, in addition to places where the FCC regulates communications?

D. From any vessel or craft located in international waters and documented or registered in the United States

T1C07 (B) [97.23]

What may result when correspondence from the FCC is returned as undeliverable because the grantee failed to provide the correct mailing address?

B. Revocation of the station license or suspension of the operator license

T1C08 (C) [97.25]

What is the normal term for an FCC-issued primary station/operator license grant?

C. Ten years

T1C09 (A) [97.21(a)(b)]

What is the grace period following the expiration of an amateur license within which the license may be renewed?

A. Two years

T1C10 (C) [97.5a]

How soon may you operate a transmitter on an amateur service frequency after you pass the examination required for your first amateur radio license?

C. As soon as your name and call sign appear in the FCC's ULS database

T1C11 (A) [97.21(b)]

If your license has expired and is still within the allowable grace period, may you continue to operate a transmitter on amateur service frequencies?

A. No, transmitting is not allowed until the ULS database shows that the license has been renewed

T1D - Authorized and prohibited transmissions

T1D01 (A) [97.111(a)(1)]

With which countries are FCC-licensed amateur stations prohibited from exchanging communications?

A. Any country whose administration has notified the ITU that it objects to such communications

T1D02 (A) [97.111(a)(5)]

On which of the following occasions may an FCC-licensed amateur station exchange messages with a U.S. military station?

A. During an Armed Forces Day Communications Test

T1D03 (C) [97.113(a)(4), 97.211(b), 97.217]

When is the transmission of codes or ciphers allowed to hide the meaning of a message transmitted by an amateur station?

C. Only when transmitting control commands to space stations or radio control craft

T1D04 (A) [97.113(a)(4), 97.113(e)]

What is the only time an amateur station is authorized to transmit music?

A. When incidental to an authorized retransmission of manned spacecraft communications

T1D05 (A) [97.113(a)(3)]

When may amateur radio operators use their stations to notify other amateurs of the availability of equipment for sale or trade?

A. When the equipment is normally used in an amateur station and such activity is not conducted on a regular basis

T1D06 (A) [97.113(a)(4)]

Which of the following types of transmissions are prohibited?

A. Transmissions that contain obscene or indecent words or language

T1D07 (B) [97.113(f)]

When is an amateur station authorized to automatically retransmit the radio signals of other amateur stations?

B. When the signals are from an auxiliary, repeater, or space station

T1D08 (B) [97.113]

When may the control operator of an amateur station receive compensation for operating the station?

B. When the communication is incidental to classroom instruction at an educational institution

T1D09 (A) [97.113(b)]

Under which of the following circumstances are amateur stations authorized to transmit signals related to broadcasting, program production, or news gathering, assuming no other means is available?

A. Only where such communications directly relate to the immediate safety of human life or protection of property

T1D10 (D) [97.3(a)(10)]

What is the meaning of the term broadcasting in the FCC rules for the amateur services?

D. Transmissions intended for reception by the general public

T1D11 (A) [97.113(a)(5)]

Which of the following types of communications are permitted in the Amateur Radio Service?

A. Brief transmissions to make station adjustments

T1E - Control operator and control types; control operator required, eligibility, designation of control operator, privileges and duties, control point, local, automatic and remote control, location of control operator

T1E01 (A) [97.7(a)]

When must an amateur station have a control operator?

A. Only when the station is transmitting

T1E02 (D) [97.7(a)]

Who is eligible to be the control operator of an amateur station?

D. Only a person for whom an amateur operator/primary station license grant appears in the FCC database or who is authorized for alien reciprocal operation

T1E03 (A) [97.103(b)]

Who must designate the station control operator?

A. The station licensee

T1E04 (D) [97.103(b)]

What determines the transmitting privileges of an amateur station?

D. The class of operator license held by the control operator

T1E05 (C) [97.3(a)(14)]

What is an amateur station control point?

C. The location at which the control operator function is performed

T1E06 (B) [97.109(d)]

Under which of the following types of control is it permissible for the control operator to be at a location other than the control point?

B. Automatic control

T1E07 (D) [97.103(a)]

When the control operator is not the station licensee, who is responsible for the proper operation of the station?

D. The control operator and the station licensee are equally responsible

T1E08 (C) [97.3(a)]

What type of control is being used for a repeater when the control operator is not present at a control point?

C. Automatic control

T1E09 (D) [97.109(a)]

What type of control is being used when transmitting using a handheld radio?

D. Local control

T1E10 (B) [97.3]

What type of control is used when the control operator is not at the station location but can indirectly manipulate the operating adjustments of a station?

B. Remote

T1E11 (D) [97.103(a)]

Who does the FCC presume to be the control operator of an amateur station, unless documentation to the contrary is in the station records?

D. The station licensee

T1F - Station identification and operation standards; special operations for repeaters and auxiliary stations, third party communications, club stations, station security, FCC inspection

T1F01 (A)

What type of identification is being used when identifying a station on the air as "Race Headquarters"?

A. Tactical call

T1F02 (C) [97.119 (a)]

When using tactical identifiers, how often must your station transmit the station's FCC-assigned call sign?

C. Every ten minutes

T1F03 (D) [97.119(a)]

When is an amateur station required to transmit its assigned call sign?

D. At least every 10 minutes during and at the end of a contact

T1F04 (C) [97.119(b)]
Which of the following is an acceptable language for use for station identification when operating in *a* phone sub-band?
C. The English language

T1F05 (B) [97.119(b)]

What method of call sign identification is required for a station transmitting phone signals?

B. Send the call sign using CW or phone emission

T1F06 (D) [97.119(c)]

Which of the following formats of a self-assigned indicator is acceptable when identifying using a phone transmission?

D. All of these choices are correct

  KL7CC stroke W3

  KL7CC slant W3

  KL7CC slash W3

T1F07 (D) [97.119(c)]

Which of the following restrictions apply when appending a self-assigned call sign indicator?

D. It must not conflict with any other indicator specified by the FCC rules or with any call sign prefix assigned to another country

T1F08 (A) [97.119(e)]

When may a Technician Class licensee be the control operator of a station operating in an exclusive Extra Class operator segment of the amateur bands?

A. Never

T1F09 (C) [97.3(a)(39)]

What type of amateur station simultaneously retransmits the signal of another amateur station on a different channel or channels?

C. Repeater station

T1F10 (A) [97.205(g)]

Who is accountable should a repeater inadvertently retransmit communications that violate the FCC rules?

A. The control operator of the originating station

T1F11 (A) [97.115(a)]

To which foreign stations do the FCC rules authorize the transmission of non-emergency third party communications?

A. Any station whose government permits such communications

T1F12 (B) [97.5(b)(2)]

How many persons are required to be members of a club for a club station license to be issued by the FCC?

B. At least 4

T1F13 (B) [97.103(c)]

When must the station licensee make the station and its records available for FCC inspection?

B. Any time upon request by an FCC representative

SUBELEMENT T2 - Operating Procedures [3 Exam Questions - 3 Groups]

T2A - Station operation; choosing an operating frequency, calling another station, test transmissions, use of minimum power, frequency use, band plans

T2A01 (B)

What is the most common repeater frequency offset in the 2 meter band?

B. plus or minus 600 kHz

T2A02 (D)

What is the national calling frequency for FM simplex operations in the 70 cm band?

D. 446.000 MHz

T2A03 (A)

What is a common repeater frequency offset in the 70 cm band?

A. Plus or minus 5 MHz

T2A04 (B)

What is an appropriate way to call another station on a repeater if you know the other station's call sign?

B. Say the station's call sign then identify with your call sign

T2A05 (C)

What should you transmit when responding to a call of CQ?

C. The other station's call sign followed by your call sign

T2A06 (A)

What must an amateur operator do when making on-air transmissions to test equipment or antennas?

A. Properly identify the transmitting station

T2A07 (D)

Which of the following is true when making a test transmission?

D. Station identification is required at least every ten minutes during the test and at the end

T2A08 (D)

What is the meaning of the procedural signal "CQ"?

D. Calling any station

T2A09 (B)

What brief statement is often used in place of "CQ" to indicate that you are listening on a repeater?

B. Say your call sign

T2A10 (A)

What is a band plan, beyond the privileges established by the FCC?

A. A voluntary guideline for using different modes or activities within an amateur band

T2A11 (D) [97.313(a)]

What are the FCC rules regarding power levels used in the amateur bands?

D. An amateur must use the minimum transmitter power necessary to carry out the desired communication

T2B – VHF/UHF operating practices; SSB phone, FM repeater, simplex, frequency offsets, splits and shifts, CTCSS, DTMF, tone squelch, carrier squelch, phonetics

T2B01 (C)

What is the term used to describe an amateur station that is transmitting and receiving on the same frequency?

C. Simplex communication

T2B02 (D)

What is the term used to describe the use of a sub-audible tone transmitted with normal voice audio to open the squelch of a receiver?

D. CTCSS

T2B03 (B)

Which of the following describes the muting of receiver audio controlled solely by the presence or absence of an RF signal?

B. Carrier squelch

T2B04 (D)

Which of the following common problems might cause you to be able to hear but not access a repeater even when transmitting with the proper offset?

A. The repeater receiver requires audio tone burst for access

B. The repeater receiver requires a CTCSS tone for access

C. The repeater receiver may require a DCS tone sequence for access

D. All of these choices are correct

T2B05 (C)

What determines the amount of deviation of an FM signal?

C. The amplitude of the modulating signal

T2B06 (A)

What happens when the deviation of an FM transmitter is increased?

A. Its signal occupies more bandwidth

T2B07 (D)

What should you do if you receive a report that your station's transmissions are causing splatter or interference on nearby frequencies?

D. Check your transmitter for off-frequency operation or spurious emissions

T2B08 (B)

What is the proper course of action if your station's transmission unintentionally interferes with another station?

B. Properly identify your transmission and move to a different frequency

T2B09 (A) [97.119(b)(2)]

Which of the following methods is encouraged by the FCC when identifying your station when using phone?

A. Use of a phonetic alphabet

T2B10 (A)

What is the "Q" signal used to indicate that you are receiving interference from other stations?

A. QRM

T2B11 (B)

What is the "Q" signal used to indicate that you are changing frequency?

B. QSY

T2C – Public service; emergency and non-emergency operations, message traffic handling

T2C01 (C) [97.103(a)]

What set of rules applies to proper operation of your station when using amateur radio at the request of public service officials?

C. FCC Rules

T2C02 (D) [97.113 and FCC Public Notice DA 09-2259]

Who must submit the request for a temporary waiver of Part 97.113 to allow amateur radio operators to provide communications on behalf of their employers during a government sponsored disaster drill?

D. The government agency sponsoring the event

T2C03 (C) [97.113]

When is it legal for an amateur licensee to provide communications on behalf of their employer during a government sponsored disaster drill or exercise?

C. Only when the FCC has granted a government-requested waiver

T2C04 (D)

What do RACES and ARES have in common?

D. Both organizations may provide communications during emergencies

T2C05 (B) [97.3(a)(37), 97.407]

What is the Radio Amateur Civil Emergency Service?

B. A radio service using amateur stations for emergency management or civil defense communications

T2C06 (C)

Which of the following is common practice during net operations to get the immediate attention of the net control station when reporting an emergency?

C. Begin your transmission with "Priority" or "Emergency" followed by your call sign

T2C07 (C)

What should you do to minimize disruptions to an emergency traffic net once you have checked in?

C. Do not transmit on the net frequency until asked to do so by the net control station

T2C08 (A)

What is usually considered to be the most important job of an amateur operator when handling emergency traffic messages?

A. Passing messages exactly as written, spoken or as received

T2C09 (B) [97.403]

When may an amateur station use any means of radio communications at its disposal for essential communications in connection with immediate safety of human life and protection of property?

B. When normal communications systems are not available

T2C10 (D)

What is the preamble in a formal traffic message?

D. The information needed to track the message as it passes through the amateur radio traffic handling system

T2C11 (A)

What is meant by the term "check" in reference to a formal traffic message?

A. The check is a count of the number of words or word equivalents in the text portion of the message

SUBELEMENT T3 – Radio wave characteristics, radio and electromagnetic properties, propagation modes – [3 Exam Questions - 3 Groups]

T3A - Radio wave characteristics; how a radio signal travels; distinctions of HF, VHF and UHF; fading, multipath; wavelength vs. penetration; antenna orientation

T3A01 (D)

What should you do if another operator reports that your station's 2 meter signals were strong just a moment ago, but now they are weak or distorted?

D. Try moving a few feet, as random reflections may be causing multi-path distortion

T3A02 (B)

Why are UHF signals often more effective from inside buildings than VHF signals?

B. The shorter wavelength allows them to more easily penetrate the structure of buildings

T3A03 (C)

What antenna polarization is normally used for long-distance weak-signal CW and SSB contacts using the VHF and UHF bands?

C. Horizontal

T3A04 (B)

What can happen if the antennas at opposite ends of a VHF or UHF line of sight radio link are not using the same polarization?

B. Signals could be significantly weaker

T3A05 (B)

When using a directional antenna, how might your station be able to access a distant repeater if buildings or obstructions are blocking the direct line of sight path?

B. Try to find a path that reflects signals to the repeater

T3A06 (B)

What term is commonly used to describe the rapid fluttering sound sometimes heard from mobile stations that are moving while transmitting?

B. Picket fencing

T3A07 (A)

What type of wave carries radio signals between transmitting and receiving stations?

A. Electromagnetic

T3A08 (C)

What is the cause of irregular fading of signals from distant stations during times of generally good reception?

C. Random combining of signals arriving via different path lengths

T3A09 (B)

Which of the following is a common effect of "skip" reflections between the Earth and the ionosphere?

B. The polarization of the original signal is randomized

T3A10 (D)

What may occur if VHF or UHF data signals propagate over multiple paths?

D. Error rates are likely to increase

T3A11 (C)

Which part of the atmosphere enables the propagation of radio signals around the world?

C. The ionosphere

T3B - Radio and electromagnetic wave properties; the electromagnetic spectrum, wavelength vs. frequency, velocity of electromagnetic waves

T3B01 (C)

What is the name for the distance a radio wave travels during one complete cycle?

C. Wavelength

T3B02 (D)

What term describes the number of times per second that an alternating current reverses direction?

D. Frequency

T3B03 (C)

What are the two components of a radio wave?

C. Electric and magnetic fields

T3B04 (A)

How fast does a radio wave travel through free space?

A. At the speed of light

T3B05 (B)

How does the wavelength of a radio wave relate to its frequency?

B. The wavelength gets shorter as the frequency increases

T3B06 (D)

What is the formula for converting frequency to wavelength in meters?

D. Wavelength in meters equals 300 divided by frequency in megahertz

T3B07 (A)

What property of radio waves is often used to identify the different frequency bands?

A. The approximate wavelength

T3B08 (B)

What are the frequency limits of the VHF spectrum?

B. 30 to 300 MHz

T3B09 (D)

What are the frequency limits of the UHF spectrum?

D. 300 to 3000 MHz

T3B10 (C)

What frequency range is referred to as HF?

C. 3 to 30 MHz

T3B11 (B)

What is the approximate velocity of a radio wave as it travels through free space?

B. 300,000,000 meters per second

T3C - Propagation modes; line of sight, sporadic E, meteor, aurora scatter, tropospheric ducting, F layer skip, radio horizon

T3C01 (C)

Why are "direct" (not via a repeater) UHF signals rarely heard from stations outside your local coverage area?

C. UHF signals are usually not reflected by the ionosphere

T3C02 (D)

Which of the following might be happening when VHF signals are being received from long distances?

D. Signals are being refracted from a sporadic E layer

T3C03 (B)

What is a characteristic of VHF signals received via auroral reflection?

B. The signals exhibit rapid fluctuations of strength and often sound distorted

T3C04 (B)

Which of the following propagation types is most commonly associated with occasional strong over-the-horizon signals on the 10, 6, and 2 meter bands?

B. Sporadic E

T3C05 (C)

What is meant by the term "knife-edge" propagation?

C. Signals are partially refracted around solid objects exhibiting sharp edges

T3C06 (A)

What mode is responsible for allowing over-the-horizon VHF and UHF communications to ranges of approximately 300 miles on a regular basis?

A. Tropospheric scatter

T3C07 (B)

What band is best suited to communicating via meteor scatter?

B. 6 meters

T3C08 (D)

What causes "tropospheric ducting"?

D. Temperature inversions in the atmosphere

T3C09 (A)

What is generally the best time for long-distance 10 meter band propagation?

A. During daylight hours

T3C10 (A)

What is the radio horizon?

A. The distance at which radio signals between two points are effectively blocked by the curvature of the Earth

T3C11 (C)

Why do VHF and UHF radio signals usually travel somewhat farther than the visual line of sight distance between two stations?

C. The Earth seems less curved to radio waves than to light

SUBELEMENT T4 - Amateur radio practices and station set up – [2 Exam Questions - 2 Groups]

T4A – Station setup; microphone, speaker, headphones, filters, power source, connecting a computer, RF grounding

T4A01 (B)

Which of the following is true concerning the microphone connectors on amateur transceivers?

B. Some connectors include push-to-talk and voltages for powering the microphone

T4A02 (C)

What could be used in place of a regular speaker to help you copy signals in a noisy area?

C. A set of headphones

T4A03 (A)

Which is a good reason to use a regulated power supply for communications equipment?

A. It prevents voltage fluctuations from reaching sensitive circuits

T4A04 (A)

Where must a filter be installed to reduce harmonic emissions?

A. Between the transmitter and the antenna

T4A05 (D)

What type of filter should be connected to a TV receiver as the first step in trying to prevent RF overload from a nearby 2 meter transmitter?

D. Band-reject filter

T4A06 (C)

Which of the following would be connected between a transceiver and computer in a packet radio station?

C. Terminal node controller

T4A07 (C)

How is the computer's sound card used when conducting digital communications using a computer?

C. The sound card provides audio to the microphone input and converts received audio to digital form

T4A08 (D)

Which type of conductor is best to use for RF grounding?

D. Flat strap

T4A09 (D)

Which would you use to reduce RF current flowing on the shield of an audio cable?

D. Ferrite choke

T4A10 (B)

What is the source of a high-pitched whine that varies with engine speed in a mobile transceiver's receive audio?

B. The alternator

T4A11 (A)

Where should a mobile transceiver's power negative connection be made?

A. At the battery or engine block ground strap

T4B - Operating controls; tuning, use of filters, squelch, AGC, repeater offset, memory channels

T4B01 (B)

What may happen if a transmitter is operated with the microphone gain set too high?

B. The output signal might become distorted

T4B02 (A)

Which of the following can be used to enter the operating frequency on a modern transceiver?

A. The keypad or VFO knob

T4B03 (D)

What is the purpose of the squelch control on a transceiver?

D. To mute receiver output noise when no signal is being received

T4B04 (B)

What is a way to enable quick access to a favorite frequency on your transceiver?

B. Store the frequency in a memory channel

T4B05 (C)

Which of the following would reduce ignition interference to a receiver?

C. Turn on the noise blanker

T4B06 (D)

Which of the following controls could be used if the voice pitch of a single-sideband signal seems too high or low?

D. The receiver RIT or clarifier

T4B07 (B)

What does the term "RIT" mean?

B. Receiver Incremental Tuning

T4B08 (B)

What is the advantage of having multiple receive bandwidth choices on a multimode transceiver?

B. Permits noise or interference reduction by selecting a bandwidth matching the mode

T4B09 (C)

Which of the following is an appropriate receive filter to select in order to minimize noise and interference for SSB reception?

C. 2400 Hz

T4B10 (A)

Which of the following is an appropriate receive filter to select in order to minimize noise and interference for CW reception?

A. 500 Hz

T4B11 (C)

Which of the following describes the common meaning of the term "repeater offset"?

C. The difference between the repeater's transmit and receive frequencies

SUBELEMENT T5 – Electrical principles, math for electronics, electronic principles, Ohm's Law – [4 Exam Questions - 4 Groups]

T5A - Electrical principles; current and voltage, conductors and insulators, alternating and direct current

T5A01 (D)

Electrical current is measured in which of the following units?

D. Amperes

T5A02 (B)

Electrical power is measured in which of the following units?

B. Watts

T5A03 (D)

What is the name for the flow of electrons in an electric circuit?

D. Current

T5A04 (B)

What is the name for a current that flows only in one direction?

B. Direct current

T5A05 (A)

What is the electrical term for the electromotive force (EMF) that causes electron flow?

A. Voltage

T5A06 (A)

How much voltage does a mobile transceiver usually require?

A. About 12 volts

T5A07 (C)

Which of the following is a good electrical conductor?

C. Copper

T5A08 (B)

Which of the following is a good electrical insulator?

B. Glass

T5A09 (A)

What is the name for a current that reverses direction on a regular basis?

A. Alternating current

T5A10 (C)

Which term describes the rate at which electrical energy is used?

C. Power

T5A11 (A)

What is the basic unit of electromotive force?

A. The volt

T5B - Math for electronics; decibels, electrical units and the metric system

T5B01 (C)

How many milliamperes is 1.5 amperes?

C. 1,500 milliamperes

T5B02 (A)

What is another way to specify a radio signal frequency of 1,500,000 hertz?

A. 1500 kHz

T5B03 (C)

How many volts are equal to one kilovolt?

C. One thousand volts

T5B04 (A)

How many volts are equal to one microvolt?

A. One one-millionth of a volt

T5B05 (B)

Which of the following is equivalent to 500 milliwatts?

B. 0.5 watts

T5B06 (C)

If an ammeter calibrated in amperes is used to measure a 3000-milliampere current, what reading would it show?

C. 3 amperes

T5B07 (C)

If a frequency readout calibrated in megahertz shows a reading of 3.525 MHz, what would it show if it were calibrated in kilohertz?

C. 3525 kHz

T5B08 (B)

How many microfarads are 1,000,000 picofarads?

B. 1 microfarad

T5B09 (B)

What is the approximate amount of change, measured in decibels (dB), of a power increase from 5 watts to 10 watts?

B. 3 dB

T5B10 (C)

What is the approximate amount of change, measured in decibels (dB), of a power decrease from 12 watts to 3 watts?

C. 6 dB

T5B11 (A)

What is the approximate amount of change, measured in decibels (dB), of a power increase from 20 watts to 200 watts?

A. 10 dB

T5C - Electronic principles; capacitance, inductance, current flow in circuits, alternating current, definition of RF, power calculations

T5C01 (D)

What is the ability to store energy in an electric field called?

D. Capacitance

T5C02 (A)

What is the basic unit of capacitance?

A. The farad

T5C03 (D)

What is the ability to store energy in a magnetic field called?

D. Inductance

T5C04 (C)

What is the basic unit of inductance?

C. The henry

T5C05 (A)

What is the unit of frequency?

A. Hertz

T5C06 (C)

What is the abbreviation that refers to radio frequency signals of all types?

C. RF

T5C07 (C)

What is a usual name for electromagnetic waves that travel through space?

C. Radio waves

T5C08 (A)

What is the formula used to calculate electrical power in a DC circuit?

A. Power (P) equals voltage (E) multiplied by current (I)

T5C09 (A)

How much power is being used in a circuit when the applied voltage is 13.8 volts DC and the current is 10 amperes?

A. 138 watts

T5C10 (B)

How much power is being used in a circuit when the applied voltage is 12 volts DC and the current is 2.5 amperes?

B. 30 watts

T5C11 (B)

How many amperes are flowing in a circuit when the applied voltage is 12 volts DC and the load is 120 watts?

B. 10 amperes

T5D – Ohm's Law

T5D01 (B)

What formula is used to calculate current in a circuit?

B. Current (I) equals voltage (E) divided by resistance (R)

T5D02 (A)

What formula is used to calculate voltage in a circuit?

A. Voltage (E) equals current (I) multiplied by resistance (R)

T5D03 (B)

What formula is used to calculate resistance in a circuit?

B. Resistance (R) equals voltage (E) divided by current (I)

T5D04 (B)

What is the resistance of a circuit in which a current of 3 amperes flows through a resistor connected to 90 volts?

B. 30 ohms

T5D05 (C)

What is the resistance in a circuit for which the applied voltage is 12 volts and the current flow is 1.5 amperes?

C. 8 ohms

T5D06 (A)

What is the resistance of a circuit that draws 4 amperes from a 12-volt source?

A. 3 ohms

T5D07 (D)

What is the current flow in a circuit with an applied voltage of 120 volts and a resistance of 80 ohms?

D. 1.5 amperes

T5D08 (C)

What is the current flowing through a 100-ohm resistor connected across 200 volts?

C. 2 amperes

T5D09 (C)

What is the current flowing through a 24-ohm resistor connected across 240 volts?

C. 10 amperes

T5D10 (A)

What is the voltage across a 2-ohm resistor if a current of 0.5 amperes flows through it?

A. 1 volt

T5D11 (B)

What is the voltage across a 10-ohm resistor if a current of 1 ampere flows through it?

B. 10 volts

T5D12 (D)

What is the voltage across a 10-ohm resistor if a current of 2 amperes flows through it?

D. 20 volts

SUBELEMENT T6 – Electrical components, semiconductors, circuit diagrams, component functions – [4 Exam Questions - 4 Groups]

T6A - Electrical components; fixed and variable resistors, capacitors, and inductors; fuses, switches, batteries

T6A01 (B)

What electrical component is used to oppose the flow of current in a DC circuit?

B. Resistor

T6A02 (C)

What type of component is often used as an adjustable volume control?

C. Potentiometer

T6A03 (B)

What electrical parameter is controlled by a potentiometer?

B. Resistance

T6A04 (B)

What electrical component stores energy in an electric field?

B. Capacitor

T6A05 (D)

What type of electrical component consists of two or more conductive surfaces separated by an insulator?

D. Capacitor

T6A06 (C)

What type of electrical component stores energy in a magnetic field?

C. Inductor

T6A07 (D)

What electrical component is usually composed of a coil of wire?

D. Inductor

T6A08 (B)

What electrical component is used to connect or disconnect electrical circuits?

B. Switch

T6A09 (A)

What electrical component is used to protect other circuit components from current overloads?

A. Fuse

T6A10 (B)

What is the nominal voltage of a fully charged nickel-cadmium cell?

B. 1.2 volts

T6A11 (B)

Which battery type is not rechargeable?

B. Carbon-zinc

T6B – Semiconductors; basic principles of diodes and transistors

T6B01 (D)

What class of electronic components is capable of using a voltage or current signal to control current flow?

D. Transistors

T6B02 (C)

What electronic component allows current to flow in only one direction?

C. Diode

T6B03 (C)

Which of these components can be used as an electronic switch or amplifier?

C. Transistor

T6B04 (B)

Which of these components is made of three layers of semiconductor material?

B. Bipolar junction transistor

T6B05 (A)

Which of the following electronic components can amplify signals?

A. Transistor

T6B06 (B)

How is a semiconductor diode's cathode lead usually identified?

B. With a stripe

T6B07 (B)

What does the abbreviation "LED" stand for?

B. Light Emitting Diode

T6B08 (A)

What does the abbreviation "FET" stand for?

A. Field Effect Transistor

T6B09 (C)

What are the names of the two electrodes of a diode?

C. Anode and cathode

T6B10 (A)

Which semiconductor component has an emitter electrode?

A. Bipolar transistor

T6B11 (B)

Which semiconductor component has a gate electrode?

B. Field effect transistor

T6B12 (A)

What is the term that describes a transistor's ability to amplify a signal?

A. Gain

T6C - Circuit diagrams; schematic symbols

T6C01 (C)

What is the name for standardized representations of components in an electrical wiring diagram?

C. Schematic symbols

T6C02 (A)

What is component 1 in figure T1?

A. Resistor

T6C03 (B)

What is component 2 in figure T1?

B. Transistor

T6C04 (C)

What is component 3 in figure T1?

C. Lamp

T6C05 (C)

What is component 4 in figure T1?

C. Battery

T6C06 (B)

What is component 6 in figure T2?

B. Capacitor

T6C07 (D)

What is component 8 in figure T2?

D. Light emitting diode

T6C08 (C)

What is component 9 in figure T2?

C. Variable resistor

T6C09 (D)

What is component 4 in figure T2?

D. Transformer

T6C10 (D)

What is component 3 in figure T3?

D. Variable inductor

T6C11 (A)

What is component 4 in figure T3?

A. Antenna

T6C12 (A)

What do the symbols on an electrical circuit schematic diagram represent?

A. Electrical components

T6C13 (C)

Which of the following is accurately represented in electrical circuit schematic diagrams?

A.   The way components are interconnected

T6D - Component functions

T6D01 (B)

Which of the following devices or circuits changes an alternating current into a varying direct current signal?

B. Rectifier

T6D02 (A)

What best describes a relay?

A. A switch controlled by an electromagnet

T6D03 (A)

What type of switch is represented by item 3 in figure T2?

A. Single-pole single-throw

T6D04 (C)

Which of the following can be used to display signal strength on a numeric scale?

C. Meter

T6D05 (A)

What type of circuit controls the amount of voltage from a power supply?

A. Regulator

T6D06 (B)

What component is commonly used to change 120V AC house current to a lower AC voltage for other uses?

B. Transformer

T6D07 (A)

Which of the following is commonly used as a visual indicator?

A. LED

T6D08 (D)

Which of the following is used together with an inductor to make a tuned circuit?

D. Capacitor

T6D09 (C)

What is the name of a device that combines several semiconductors and other components into one package?

C. Integrated circuit

T6D10 (C)

What is the function of component 2 in Figure T1?

C. Control the flow of current

T6D11 (B)

Which of the following is a common use of coaxial cable?

B. Carry RF signals between a radio and antenna

SUBELEMENT T7 – Station equipment; common transmitter and receiver problems, antenna measurements and troubleshooting, basic repair and testing – [4 Exam Questions - 4 Groups]

T7A - Station radios; receivers, transmitters, transceivers

T7A01 (C)

What is the function of a product detector?

C. Detect CW and SSB signals

T7A02 (C)

What type of receiver is shown in Figure T6?

C. Single-conversion superheterodyne

T7A03 (C)

What is the function of a mixer in a superheterodyne receiver?

C. To shift the incoming signal to an intermediate frequency

T7A04 (D)

What circuit is pictured in Figure T7, if block 1 is a frequency discriminator?

D. An FM receiver

T7A05 (D)

What is the function of block 1 if figure T4 is a simple CW transmitter?

D. Oscillator

T7A06 (C)

What device takes the output of a low-powered 28 MHz SSB exciter and produces a 222 MHz output signal?

C. Transverter

T7A07 (B)

If figure T5 represents a transceiver in which block 1 is the transmitter portion and block 3 is the receiver portion, what is the function of block 2?

B. A transmit-receive switch

T7A08 (C)

Which of the following circuits combines a speech signal and an RF carrier?

C. Modulator

T7A09 (B)

Which of the following devices is most useful for VHF weak-signal communication?

B. A multi-mode VHF transceiver

T7A10 (B)

What device increases the low-power output from a handheld transceiver?

B. An RF power amplifier

T7A11 (B)

Which of the following circuits demodulates FM signals?

B. Discriminator

T7A12 (C)

Which term describes the ability of a receiver to discriminate between multiple signals?

C. Selectivity

T7A13 (A)

Where is an RF preamplifier installed?

A. Between the antenna and receiver

T7B – Common transmitter and receiver problems; symptoms of overload and overdrive, distortion, interference, over and under modulation, RF feedback, off frequency signals; fading and noise; problems with digital communications interfaces

T7B01 (D)

What can you do if you are told your FM handheld or mobile transceiver is over deviating?

D. Talk farther away from the microphone

T7B02 (C)

What is meant by fundamental overload in reference to a receiver?

C. Interference caused by very strong signals

T7B03 (D)

Which of the following may be a cause of radio frequency interference?

A. Fundamental overload

B. Harmonics

C. Spurious emissions

D. All of these choices are correct

T7B04 (B)

What is the most likely cause of interference to a non-cordless telephone from a nearby transmitter?

B. The telephone is inadvertently acting as a radio receiver

T7B05 (C)

What is a logical first step when attempting to cure a radio frequency interference problem in a nearby telephone?

C. Install an RF filter at the telephone

T7B06 (A)

What should you do first if someone tells you that your station's transmissions are interfering with their radio or TV reception?

A. Make sure that your station is functioning properly and that it does not cause interference to your own television

T7B07 (D)

Which of the following may be useful in correcting a radio frequency interference problem?

A. Snap-on ferrite chokes

B. Low-pass and high-pass filters

C. Band-reject and band-pass filters

D. All of these choices are correct

T7B08 (D)

What should you do if a "Part 15" device in your neighbor's home is causing harmful interference to your amateur station?

A. Work with your neighbor to identify the offending device

B. Politely inform your neighbor about the rules that require him to stop using the device if it causes interference

C. Check your station and make sure it meets the standards of good amateur practice

D. All of these choices are correct

T7B09 (D)

What could be happening if another operator reports a variable high-pitched whine on the audio from your mobile transmitter?

D. Noise on the vehicle's electrical system is being transmitted along with your speech audio

T7B10 (D)

What might be the problem if you receive a report that your audio signal through the repeater is distorted or unintelligible?

A. Your transmitter may be slightly off frequency

B. Your batteries may be running low

C. You could be in a bad location

D. All of these choices are correct

T7B11 (C)

What is a symptom of RF feedback in a transmitter or transceiver?

C. Reports of garbled, distorted, or unintelligible transmissions

T7B12 (C)

What does the acronym "BER" mean when applied to digital communications systems?

C. Bit Error Rate

T7C – Antenna measurements and troubleshooting; measuring SWR, dummy loads, feedline failure modes

T7C01 (A)

What is the primary purpose of a dummy load?

A. To prevent the radiation of signals when making tests

T7C02 (B)

Which of the following instruments can be used to determine if an antenna is resonant at the desired operating frequency?

B. An antenna analyzer

T7C03 (A)

What, in general terms, is standing wave ratio (SWR)?

A. A measure of how well a load is matched to a transmission line

T7C04 (C)

What reading on an SWR meter indicates a perfect impedance match between the antenna and the feedline?

C. 1 to 1

T7C05 (A)

What is the approximate SWR value above which the protection circuits in most solid-state transmitters begin to reduce transmitter power?

A. 2 to 1

T7C06 (D)

What does an SWR reading of 4:1 mean?

D. An impedance mismatch

T7C07 (C)

What happens to power lost in a feedline?

C. It is converted into heat

T7C08 (D)

What instrument other than an SWR meter could you use to determine if a feedline and antenna are properly matched?

D. Directional wattmeter

T7C09 (A)

Which of the following is the most common cause for failure of coaxial cables?

A. Moisture contamination

T7C10 (D)

Why should the outer jacket of coaxial cable be resistant to ultraviolet light?

D. Ultraviolet light can damage the jacket and allow water to enter the cable

T7C11 (C)

What is a disadvantage of "air core" coaxial cable when compared to foam or solid dielectric types?

C. It requires special techniques to prevent water absorption

T7D – Basic repair and testing; soldering, use of a voltmeter, ammeter, and ohmmeter

T7D01 (B)

Which instrument would you use to measure electric potential or electromotive force?

B. A voltmeter

T7D02 (B)

What is the correct way to connect a voltmeter to a circuit?

B. In parallel with the circuit

T7D03 (A)

How is an ammeter usually connected to a circuit?

A. In series with the circuit

T7D04 (D)

Which instrument is used to measure electric current?

D. An ammeter

T7D05 (D)

What instrument is used to measure resistance?

D. An ohmmeter

T7D06 (C)

Which of the following might damage a multimeter?

C. Attempting to measure voltage when using the resistance setting

T7D07 (D)

Which of the following measurements are commonly made using a multimeter?

D. Voltage and resistance

T7D08 (C)

Which of the following types of solder is best for radio and electronic use?

C. Rosin-core solder

T7D09 (C)

What is the characteristic appearance of a "cold" solder joint?

C. A grainy or dull surface

T7D10 (B)

What is probably happening when an ohmmeter, connected across a circuit, initially indicates a low resistance and then shows increasing resistance with time?

B. The circuit contains a large capacitor

T7D11 (B)

Which of the following precautions should be taken when measuring circuit resistance with an ohmmeter?

B. Ensure that the circuit is not powered

SUBELEMENT T8 – Modulation modes; amateur satellite operation, operating activities, non-voice communications – [4 Exam Questions - 4 Groups]

T8A – Modulation modes; bandwidth of various signals

T8A01 (C)

Which of the following is a form of amplitude modulation?

C. Single sideband

T8A02 (A)

What type of modulation is most commonly used for VHF packet radio transmissions?

A. FM

T8A03 (C)

Which type of voice modulation is most often used for long-distance or weak signal contacts on the VHF and UHF bands?

C. SSB

T8A04 (D)

Which type of modulation is most commonly used for VHF and UHF voice repeaters?

D. FM

T8A05 (C)

Which of the following types of emission has the narrowest bandwidth?

C. CW

T8A06 (A)

Which sideband is normally used for 10 meter HF, VHF and UHF single-sideband communications?

A. Upper sideband

T8A07 (C)

What is the primary advantage of single sideband over FM for voice transmissions?

C. SSB signals have narrower bandwidth

T8A08 (B)

What is the approximate bandwidth of a single sideband voice signal?

B. 3 kHz

T8A09 (C)

What is the approximate bandwidth of a VHF repeater FM phone signal?

C. Between 5 and 15 kHz

T8A10 (B)

What is the typical bandwidth of analog fast-scan TV transmissions on the 70 cm band?

B. About 6 MHz

T8A11 (B)

What is the approximate maximum bandwidth required to transmit a CW signal?

B. 150 Hz

T8B - Amateur satellite operation; Doppler shift, basic orbits, operating protocols

T8B01 (D)

Who may be the control operator of a station communicating through an amateur satellite or space station?

D. Any amateur whose license privileges allow them to transmit on the satellite uplink frequency

T8B02 (B) [97.313(a)]

How much transmitter power should be used on the uplink frequency of an amateur satellite or space station?

B. The minimum amount of power needed to complete the contact

T8B03 (A)

Which of the following can be done using an amateur radio satellite?

A. Talk to amateur radio operators in other countries

T8B04 (B)

Which amateur stations may make contact with an amateur station on the International Space Station using 2 meter and 70 cm band amateur radio frequencies?

B. Any amateur holding a Technician or higher class license

T8B05 (D)

What is a satellite beacon?

D. A transmission from a space station that contains information about a satellite

T8B06 (D)

What can be used to determine the time period during which an amateur satellite or space station can be accessed?

D. A satellite tracking program

T8B07 (C)

With regard to satellite communications, what is Doppler shift?

C. An observed change in signal frequency caused by relative motion between the satellite and the earth station

T8B08 (B)

What is meant by the statement that a satellite is operating in "mode U/V"?

B. The satellite uplink is in the 70 cm band and the downlink is in the 2 meter band

T8B09 (B)

What causes "spin fading" when referring to satellite signals?

B. Rotation of the satellite and its antennas

T8B10 (C)

What do the initials LEO tell you about an amateur satellite?

C. The satellite is in a Low Earth Orbit

T8B11 (C)

What is a commonly used method of sending signals to and from a digital satellite?

C. FM Packet

T8C – Operating activities; radio direction finding, radio control, contests, special event stations, basic linking over Internet

T8C01 (C)

Which of the following methods is used to locate sources of noise interference or jamming?

C. Radio direction finding

T8C02 (B)

Which of these items would be useful for a hidden transmitter hunt?

B. A directional antenna

T8C03 (A)

What popular operating activity involves contacting as many stations as possible during a specified period of time?

A. Contesting

T8C04 (C)

Which of the following is good procedure when contacting another station in a radio contest?

C. Send only the minimum information needed for proper identification and the contest exchange

T8C05 (A)

What is a grid locator?

A. A letter-number designator assigned to a geographic location

T8C06 (C)

For what purpose is a temporary "1 by 1" format (letter-number-letter) call sign assigned?

C. For operations in conjunction with an activity of special significance to the amateur community

T8C07 (B) [97.215(c)]

What is the maximum power allowed when transmitting telecommand signals to radio controlled models?

B. 1 watt

T8C08 (C) [97.215(a)]

What is required in place of on-air station identification when sending signals to a radio control model using amateur frequencies?

C. A label indicating the licensee's name, call sign and address must be affixed to the transmitter

T8C09 (C)

How might you obtain a list of active nodes that use VoIP?

C. From a repeater directory

T8C10 (D)

How do you select a specific IRLP node when using a portable transceiver?

D. Use the keypad to transmit the IRLP node ID

T8C11 (A)

What name is given to an amateur radio station that is used to connect other amateur stations to the Internet?

A. A gateway

T8D – Non-voice communications; image data, digital modes, CW, packet, PSK31

T8D01 (D)

Which of the following is an example of a digital communications method?

A. Packet

B. PSK31

C. MFSK

D. All of these choices are correct

T8D02 (A)

What does the term APRS mean?

A. Automatic Position Reporting System

T8D03 (D)

Which of the following is normally used when sending automatic location reports via amateur radio?

D. A Global Positioning System receiver

T8D04 (C)

What type of transmission is indicated by the term NTSC?

C. An analog fast scan color TV signal

T8D05 (B)

Which of the following emission modes may be used by a Technician Class operator between 219 and 220 MHz?

B. Data

T8D06 (B)

What does the abbreviation PSK mean?

B. Phase Shift Keying

T8D07 (D)

What is PSK31?

D. A low-rate data transmission mode

T8D08 (D)

Which of the following may be included in packet transmissions?

A. A check sum which permits error detection

B. A header which contains the call sign of the station to which the information is being sent

C. Automatic repeat request in case of error

D. All of these choices are correct

T8D09 (C)

What code is used when sending CW in the amateur bands?

C. International Morse

T8D10 (D)

Which of the following can be used to transmit CW in the amateur bands?

A. Straight Key

B. Electronic Keyer

C. Computer Keyboard

D. All of these choices are correct

T8D11 (C)

What is a "parity" bit?

C. An extra code element used to detect errors in received data

SUBELEMENT T9 – Antennas, feedlines - [2 Exam Questions - 2 Groups]

T9A – Antennas; vertical and horizontal, concept of gain, common portable and mobile antennas, relationships between antenna length and frequency

T9A01 (C)

What is a beam antenna?

C. An antenna that concentrates signals in one direction

T9A02 (B)

Which of the following is true regarding vertical antennas?

B. The electric field is perpendicular to the Earth

T9A03 (B)

Which of the following describes a simple dipole mounted so the conductor is parallel to the Earth's surface?

B. A horizontally polarized antenna

T9A04 (A)

What is a disadvantage of the "rubber duck" antenna supplied with most handheld radio transceivers?

A. It does not transmit or receive as effectively as a full-sized antenna

T9A05 (C)

How would you change a dipole antenna to make it resonant on a higher frequency?

C. Shorten it

T9A06 (C)

What type of antennas are the quad, Yagi, and dish?

C. Directional antennas

T9A07 (A)

What is a good reason not to use a "rubber duck" antenna inside your car?

A. Signals can be significantly weaker than when it is outside of the vehicle

T9A08 (C)

What is the approximate length, in inches, of a quarter-wavelength vertical antenna for 146 MHz?

C. 19

T9A09 (C)

What is the approximate length, in inches, of a 6 meter 1/2-wavelength wire dipole antenna?

C. 112

T9A10 (C)

In which direction is the radiation strongest from a half-wave dipole antenna in free space?

C. Broadside to the antenna

T9A11 (C)

What is meant by the gain of an antenna?

C. The increase in signal strength in a specified direction when compared to a reference antenna

T9B - Feedlines; types, losses vs. frequency, SWR concepts, matching weather protection, connectors

T9B01 (B)

Why is it important to have a low SWR in an antenna system that uses coaxial cable feedline?

B. To allow the efficient transfer of power and reduce losses

T9B02 (B)

What is the impedance of the most commonly used coaxial cable in typical amateur radio installations?

B. 50 ohms

T9B03 (A)

Why is coaxial cable used more often than any other feedline for amateur radio antenna systems?

A. It is easy to use and requires few special installation considerations

T9B04 (A)

What does an antenna tuner do?

A. It matches the antenna system impedance to the transceiver's output impedance

T9B05 (D)

What generally happens as the frequency of a signal passing through coaxial cable is increased?

D. The loss increases

T9B06 (B)

Which of the following connectors is most suitable for frequencies above 400 MHz?

B. A Type N connector

T9B07 (C)

Which of the following is true of PL-259 type coax connectors?

C. The are commonly used at HF frequencies

T9B08 (A)

Why should coax connectors exposed to the weather be sealed against water intrusion?

A. To prevent an increase in feedline loss

T9B09 (B)

What might cause erratic changes in SWR readings?

B. A loose connection in an antenna or a feedline

T9B10 (C)

What electrical difference exists between the smaller RG-58 and larger RG-8 coaxial cables?

C. RG-8 cable has less loss at a given frequency

T9B11 (C)

Which of the following types of feedline has the lowest loss at VHF and UHF?

C. Air-insulated hard line

SUBELEMENT T0 – AC power circuits, antenna installation, RF hazards – [3 Exam Questions - 3 Groups]

T0A – AC power circuits; hazardous voltages, fuses and circuit breakers, grounding, lightning protection, battery safety, electrical code compliance

T0A01 (B)

Which is a commonly accepted value for the lowest voltage that can cause a dangerous electric shock?

B. 30 volts

T0A02 (D)

How does current flowing through the body cause a health hazard?

A. By heating tissue

B. It disrupts the electrical functions of cells

C. It causes involuntary muscle contractions

D. All of these choices are correct

T0A03 (C)

What is connected to the green wire in a three-wire electrical AC plug?

C. Safety ground

T0A04 (B)

What is the purpose of a fuse in an electrical circuit?

B. To interrupt power in case of overload

T0A05 (C)

Why is it unwise to install a 20-ampere fuse in the place of a 5-ampere fuse?

C. Excessive current could cause a fire

T0A06 (D)

What is a good way to guard against electrical shock at your station?

A. Use three-wire cords and plugs for all AC powered equipment

B. Connect all AC powered station equipment to a common safety ground

C. Use a circuit protected by a ground-fault interrupter

D. All of these choices are correct

T0A07 (D)

Which of these precautions should be taken when installing devices for lightning protection in a coaxial cable feedline?

D. Ground all of the protectors to a common plate which is in turn connected to an external ground

T0A08 (C)

What is one way to recharge a 12-volt lead-acid station battery if the commercial power is out?

C. Connect the battery to a car's battery and run the engine

T0A09 (C)

What kind of hazard is presented by a conventional 12-volt storage battery?

C. Explosive gas can collect if not properly vented

T0A10 (A)

What can happen if a lead-acid storage battery is charged or discharged too quickly?

A. The battery could overheat and give off flammable gas or explode

T0A11 (C)

Which of the following is good practice when installing ground wires on a tower for lightning protection?

C. Ensure that connections are short and direct

T0A12 (D)

What kind of hazard might exist in a power supply when it is turned off and disconnected?

D. You might receive an electric shock from stored charge in large capacitors

T0A13 (A)

What safety equipment should always be included in home-built equipment that is powered from 120V AC power circuits?

A. A fuse or circuit breaker in series with the AC "hot" conductor

T0B – Antenna installation; tower safety, overhead power lines

T0B01 (C)

When should members of a tower work team wear a hard hat and safety glasses?

C. At all times when any work is being done on the tower

T0B02 (C)

What is a good precaution to observe before climbing an antenna tower?

C. Put on a climbing harness and safety glasses

T0B03 (D)

Under what circumstances is it safe to climb a tower without a helper or observer?

D. Never

T0B04 (C)

Which of the following is an important safety precaution to observe when putting up an antenna tower?

C. Look for and stay clear of any overhead electrical wires

T0B05 (C)

What is the purpose of a gin pole?

C. To lift tower sections or antennas

T0B06 (D)

What is the minimum safe distance from a power line to allow when installing an antenna?

D. So that if the antenna falls unexpectedly, no part of it can come closer than 10 feet to the power wires

T0B07 (C)

Which of the following is an important safety rule to remember when using a crank-up tower?

C. This type of tower must never be climbed unless it is in the fully retracted position

T0B08 (C)

What is considered to be a proper grounding method for a tower?

C. Separate eight-foot long ground rods for each tower leg, bonded to the tower and each other

T0B09 (C)

Why should you avoid attaching an antenna to a utility pole?

C. The antenna could contact high-voltage power wires

T0B10 (C)

Which of the following is true concerning grounding conductors used for lightning protection?

C. Sharp bends must be avoided

T0B11 (B)

Which of the following establishes grounding requirements for an amateur radio tower or antenna?

B. Local electrical codes

T0C - RF hazards; radiation exposure, proximity to antennas, recognized safe power levels, exposure to others

T0C01 (D)

What type of radiation are VHF and UHF radio signals?

D. Non-ionizing radiation

T0C02 (B)

Which of the following frequencies has the lowest Maximum Permissible Exposure limit?

B. 50 MHz

T0C03 (C)

What is the maximum power level that an amateur radio station may use at VHF frequencies before an RF exposure evaluation is required?

C. 50 watts PEP at the antenna

T0C04 (D)

What factors affect the RF exposure of people near an amateur station antenna?

A. Frequency and power level of the RF field

B. Distance from the antenna to a person

C. Radiation pattern of the antenna

D. All of these choices are correct

T0C05 (D)

Why do exposure limits vary with frequency?

D. The human body absorbs more RF energy at some frequencies than at others

T0C06 (D)

Which of the following is an acceptable method to determine that your station complies with FCC RF exposure regulations?

A. By calculation based on FCC OET Bulletin 65

B. By calculation based on computer modeling

C. By measurement of field strength using calibrated equipment

D. All of these choices are correct

T0C07 (B)

What could happen if a person accidentally touched your antenna while you were transmitting?

B. They might receive a painful RF burn

T0C08 (A)

Which of the following actions might amateur operators take to prevent exposure to RF radiation in excess of FCC-supplied limits?

A. Relocate antennas

T0C09 (B)

How can you make sure your station stays in compliance with RF safety regulations?

B. By re-evaluating the station whenever an item of equipment is changed

T0C10 (A)

Why is duty cycle one of the factors used to determine safe RF radiation exposure levels?

A. It affects the average exposure of people to radiation

T0C11 (C)

What is meant by "duty cycle" when referring to RF exposure?

C. The ratio of on-air time to total operating time of a transmitted signal

[End of Technician Class pool. Graphics included below.]

Graphics for 2010 Technician exam - Element 2

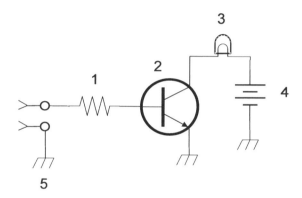

Figure T1

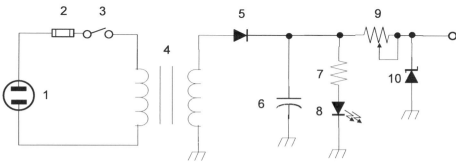

Figure T2

The Complete Study Guide For All Amateur Radio Tests

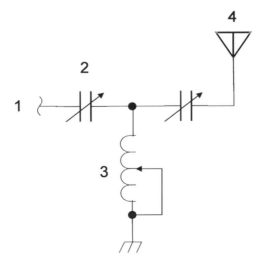

Figure T3

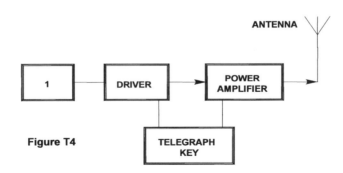

Figure T4

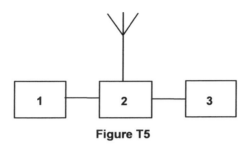

Figure T5

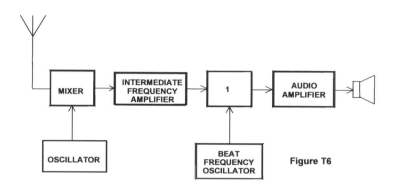

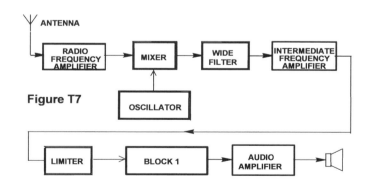

End of Technician Question Pool and Answers.

The Complete Study Guide For All Amateur Radio Tests

GENERAL CLASS

FINAL POOL RELEASE – MARCH 1, 2011

(Corrected to reflect correct answer to G7A03 per errata)

Issued by the NCVEC QPC - February 23, 2011

Edits and Corrections to the General Class, Element 3 Pool

Released December 6, 2010 to be effective July 1, 2011- June 30, 2015

We have received feedback via the NCVEC website and through direct communications pointing out minor errors and suggesting clarifications for the 2011 General Class Pool just released. The QPC has reviewed all of these comments, and is now issuing the attached minor edits and clarifications to the pool. Note that a minor error in the schematic was corrected as well.

General Class Question Pool Effective July 1, 2011 - June 30, 2015

SUBELEMENT G1 - COMMISSION'S RULES [5 Exam Questions - 5 Groups]

G1A - General Class control operator frequency privileges; primary and secondary allocations

G1A01 (C) [97.301(d), 97.303(s)]

On which of the following bands is a General Class license holder granted all amateur frequency privileges?

C. 160, 60, 30, 17, 12, and 10 meters

G1A02 (B) [97.305]

On which of the following bands is phone operation prohibited?

B. 30 meters

G1A03 (B) [97.305]

On which of the following bands is image transmission prohibited?

B. 30 meters

G1A04 (D) [97.303 (s)]

Which of the following amateur bands is restricted to communication on only specific channels, rather than frequency ranges?

D. 60 meters

G1A05 (A) [97.301(d)]

Which of the following frequencies is in the General Class portion of the 40 meter band?

A. 7.250 MHz

G1A06 (D) [97.301(d)]

Which of the following frequencies is in the 12 meter band?

D. 24.940 MHz

G1A07 (C) [97.301(d)]

Which of the following frequencies is within the General Class portion of the 75 meter phone band?

C. 3900 kHz

G1A08 (C) [97.301(d)]

Which of the following frequencies is within the General Class portion of the 20 meter phone band?

C. 14305 kHz

G1A09 (C) [97.301(d)]

Which of the following frequencies is within the General Class portion of the 80 meter band?

C. 3560 kHz

G1A10 (C) [97.301(d)]

Which of the following frequencies is within the General Class portion of the 15 meter band?

C. 21300 kHz

G1A11 (D) [97.301(d)]

Which of the following frequencies is available to a control operator holding a General Class license?

A. 28.020 MHz

B. 28.350 MHz

C. 28.550 MHz

D. All of these choices are correct

G1A12 (B) [97.301]

When General Class licensees are not permitted to use the entire voice portion of a particular band, which portion of the voice segment is generally available to them?

B. The upper frequency end

G1A13 (D) [97.303]

Which, if any, amateur band is shared with the Citizens Radio Service?

D. None

G1A14 (C) [97.303]

Which of the following applies when the FCC rules designate the Amateur Service as a secondary user on a band?

C. Amateur stations are allowed to use the band only if they do not cause harmful interference to primary users

G1A15 (D) [97.303]

What is the appropriate action if, when operating on either the 30 or 60 meter bands, a station in the primary service interferes with your contact?

D. Move to a clear frequency

G1B - Antenna structure limitations; good engineering and good amateur practice; beacon operation; restricted operation; retransmitting radio signals

G1B01 (C) [97.15(a)]

What is the maximum height above ground to which an antenna structure may be erected without requiring notification to the FAA and registration with the FCC, provided it is not at or near a public use airport?

C. 200 feet

G1B02 (D) [97.203(b)]

With which of the following conditions must beacon stations comply?

D. There must be no more than one beacon signal in the same band from a single location

G1B03 (A) [97.3(a)(9)]

Which of the following is a purpose of a beacon station as identified in the FCC Rules?

A. Observation of propagation and reception

G1B04 (A) [97.113(b)]

Which of the following must be true before amateur stations may provide communications to broadcasters for dissemination to the public?

A. The communications must directly relate to the immediate safety of human life or protection of property and there must be no other means of communication reasonably available before or at the time of the event

G1B05 (D) [97.113(a)(5),(e)]

When may music be transmitted by an amateur station?

D. When it is an incidental part of a manned space craft retransmission

G1B06 (B) [97.113(a)(4) and 97.207(f)]

When is an amateur station permitted to transmit secret codes?

B. To control a space station

G1B07 (B) [97.113(a)(4)]

What are the restrictions on the use of abbreviations or procedural signals in the Amateur Service?

B. They may be used if they do not obscure the meaning of a message

G1B08 (D)

When choosing a transmitting frequency, what should you do to comply with good amateur practice?

A. Review FCC Part 97 Rules regarding permitted frequencies and emissions?

B. Follow generally accepted band plans agreed to by the Amateur Radio community.

C. Before transmitting, listen to avoid interfering with ongoing communication

D. All of these choices are correct

G1B09 (A) [97.113(a)(3)]

When may an amateur station transmit communications in which the licensee or control operator has a pecuniary (monetary) interest?

A. When other amateurs are being notified of the sale of apparatus normally used in an amateur station and such activity is not done on a regular basis

G1B10 (C) [97.203(c)]

What is the power limit for beacon stations?

C. 100 watts PEP output

G1B11 (C) [97.101(a)]

How does the FCC require an amateur station to be operated in all respects not specifically covered by the Part 97 rules?

C. In conformance with good engineering and good amateur practice

G1B12 (A) [97.101(a)]

Who or what determines "good engineering and good amateur practice" as applied to the operation of an amateur station in all respects not covered by the Part 97 rules?

A. The FCC

G1C - Transmitter power regulations; data emission standards

G1C01 (A) [97.313(c)(1)]

What is the maximum transmitting power an amateur station may use on 10.140 MHz?

A. 200 watts PEP output

G1C02 (C) [97.313(a),(b)]

What is the maximum transmitting power an amateur station may use on the 12 meter band?

C. 1500 watts PEP output

G1C03 (A) [97.303s]

What is the maximum bandwidth permitted by FCC rules for Amateur Radio stations when transmitting on USB frequencies in the 60 meter band?

A. 2.8 kHz

G1C04 (A) [97.313]

Which of the following is a limitation on transmitter power on the 14 MHz band?

A. Only the minimum power necessary to carry out the desired communications should be used

G1C05 (C) [97.313]

Which of the following is a limitation on transmitter power on the 28 MHz band?

C. 1500 watts PEP output

G1C06 (D) [97.313]

Which of the following is a limitation on transmitter power on 1.8 MHz band?

D. 1500 watts PEP output

G1C07 (D) [97.305(c), 97.307(f)(3)]

What is the maximum symbol rate permitted for RTTY or data emission transmission on the 20 meter band?

D. 300 baud

G1C08 (D) [97.307(f)(3)]

What is the maximum symbol rate permitted for RTTY or data emission transmitted at frequencies below 28 MHz?

D. 300 baud

G1C09 (A) [97.305(c) and 97.307(f)(5)]

What is the maximum symbol rate permitted for RTTY or data emission transmitted on the 1.25 meter and 70 centimeter bands

A. 56 kilobaud

G1C10 (C) [97.305(c) and 97.307(f)(4)]

What is the maximum symbol rate permitted for RTTY or data emission transmissions on the 10 meter band?

C. 1200 baud

G1C11 (B) [97.305(c) and 97.307(f)(5)]

What is the maximum symbol rate permitted for RTTY or data emission transmissions on the 2 meter band?

B. 19.6 kilobaud

G1D - Volunteer Examiners and Volunteer Examiner Coordinators; temporary identification

G1D01 (C) [97.119(f)(2)]

Which of the following is a proper way to identify when transmitting using phone on General Class frequencies if you have a CSCE for the required elements but your upgrade from Technician has not appeared in the FCC database?

C. Give your call sign followed by "slant AG"

G1D02 (C) [97.509(b)(3)(i)]

What license examinations may you administer when you are an accredited VE holding a General Class operator license?

C. Technician only

G1D03 (C) [97.9(b)]

On which of the following band segments may you operate if you are a Technician Class operator and have a CSCE for General Class privileges?

C. On any General or Technician Class band segment

G1D04 (A) [97.509(a)(b)]

Which of the following is a requirement for administering a Technician Class operator examination?

A. At least three VEC accredited General Class or higher VEs must be present

G1D05 (D) [97.509(b)(3)(i)]

Which of the following is sufficient for you to be an administering VE for a Technician Class operator license examination?

D. An FCC General Class or higher license and VEC accreditation

G1D06 (A) [97.119(f)(2)]

When must you add the special identifier "AG" after your call sign if you are a Technician Class licensee and have a CSCE for General Class operator privileges, but the FCC has not yet posted your upgrade on its Web site?

A. Whenever you operate using General Class frequency privileges

G1D07 (C) [97.509(b)(1)]

Volunteer Examiners are accredited by what organization?

C. A Volunteer Examiner Coordinator

G1D08 (B) [97.509(b)(3)]

Which of the following criteria must be met for a non-U.S. citizen to be an accredited Volunteer Examiner?

B. The person must hold an FCC granted Amateur Radio license of General Class or above

G1D09 (C) [97.9(b)]

How long is a Certificate of Successful Completion of Examination (CSCE) valid for exam element credit?

C. 365 days

G1D10 (B) [97.509(b)(2)]

What is the minimum age that one must be to qualify as an accredited Volunteer Examiner?

B. 18 years

G1E - Control categories; repeater regulations; harmful interference; third party rules; ITU regions

G1E01 (A) [97.115(b)(2)]

Which of the following would disqualify a third party from participating in stating a message over an amateur station?

A. The third party's amateur license had ever been revoked

G1E02 (D) [97.205(a)]

When may a 10 meter repeater retransmit the 2 meter signal from a station having a Technician Class control operator?

D. Only if the 10 meter repeater control operator holds at least a General Class license

G1E03 (B) [97.301(d)]

In what ITU region is operation in the 7.175 to 7.300 MHz band permitted for a control operator holding an FCC-issued General Class license?

B. Region 2

G1E04 (D) [97.13(b),97.311(b),97.303]

Which of the following conditions require an Amateur Radio station licensee to take specific steps to avoid harmful interference to other users or facilities?

A. When operating within one mile of an FCC Monitoring Station

B. When using a band where the Amateur Service is secondary

C. When a station is transmitting spread spectrum emissions

D. All of these choices are correct

G1E05 (C) [97.115(a)(2),97.117]

What types of messages for a third party in another country may be transmitted by an amateur station?

C. Only messages relating to Amateur Radio or remarks of a personal character, or messages relating to emergencies or disaster relief

G1E06 (A) [97.205(c)]

Which of the following applies in the event of interference between a coordinated repeater and an uncoordinated repeater?

A. The licensee of the non-coordinated repeater has primary responsibility to resolve the interference

G1E07 (C) [97.115(a)(2)]

With which foreign countries is third party traffic prohibited, except for messages directly involving emergencies or disaster relief communications?

C. Every foreign country, unless there is a third party agreement in effect with that country

G1E08 (B) [97.115(a)(b)]

Which of the following is a requirement for a non-licensed person to communicate with a foreign Amateur Radio station from a station with an FCC granted license at which a licensed control operator is present?

B. The foreign amateur station must be in a country with which the United States has a third party agreement

G1E09 (C) [97.119(b)(2)]

What language must you use when identifying your station if you are using a language other than English in making a contact using phone emission?

D. Any language of a country that is a member of the ITU

G1E10 (D) [97.205(b)]

What portion of the 10 meter band is available for repeater use?

D. The portion above 29.5 MHz

SUBELEMENT G2 - OPERATING PROCEDURES [5 Exam Questions - 5 Groups]

G2A - Phone operating procedures; USB/LSB utilization conventions; procedural signals; breaking into a QSO in progress; VOX operation

G2A01 (A)

Which sideband is most commonly used for voice communications on frequencies of 14 MHz or higher?

A. Upper sideband

G2A02 (B)

Which of the following modes is most commonly used for voice communications on the 160, 75, and 40 meter bands?

B. Lower sideband

G2A03 (A)

Which of the following is most commonly used for SSB voice communications in the VHF and UHF bands?

A. Upper sideband

G2A04 (A)

Which mode is most commonly used for voice communications on the 17 and 12 meter bands?

A. Upper sideband

G2A05 (C)

Which mode of voice communication is most commonly used on the high frequency amateur bands?

C. Single sideband

G2A06 (B)

Which of the following is an advantage when using single sideband as compared to other analog voice modes on the HF amateur bands?

B. Less bandwidth used and higher power efficiency

G2A07 (B)

Which of the following statements is true of the single sideband (SSB) voice mode?

B. Only one sideband is transmitted; the other sideband and carrier are suppressed

G2A08 (B)

Which of the following is a recommended way to break into a conversation when using phone?

B. Say your call sign during a break between transmissions from the other stations

G2A09 (D)

Why do most amateur stations use lower sideband on the 160, 75 and 40 meter bands?

D. Current amateur practice is to use lower sideband on these frequency bands

G2A10 (B)

Which of the following statements is true of SSB VOX operation?

B. VOX allows "hands free" operation

G2A11 (C)

What does the expression "CQ DX" usually indicate?

C. The caller is looking for any station outside their own country

G2B - Operating courtesy; band plans; emergencies, including drills and emergency communications

G2B01 (C)

Which of the following is true concerning access to frequencies?

C. No one has priority access to frequencies, common courtesy should be a guide

G2B02 (B)

What is the first thing you should do if you are communicating with another amateur station and hear a station in distress break in?

B. Acknowledge the station in distress and determine what assistance may be needed

G2B03 (C)

If propagation changes during your contact and you notice increasing interference from other activity on the same frequency, what should you do?

C. As a common courtesy, move your contact to another frequency

G2B04 (B)

When selecting a CW transmitting frequency, what minimum frequency separation should you allow in order to minimize interference to stations on adjacent frequencies?

B. 150 to 500 Hz

G2B05 (B)

What is the customary minimum frequency separation between SSB signals under normal conditions?

B. Approximately 3 kHz

G2B06 (A)

What is a practical way to avoid harmful interference when selecting a frequency to call CQ on CW or phone?

A. Send "QRL?" on CW, followed by your call sign; or, if using phone, ask if the frequency is in use, followed by your call sign

G2B07 (C)

Which of the following complies with good amateur practice when choosing a frequency on which to initiate a call?

C. Follow the voluntary band plan for the operating mode you intend to use

G2B08 (A)

What is the "DX window" in a voluntary band plan?

A. A portion of the band that should not be used for contacts between stations within the 48 contiguous United States

G2B09 (A) [97.407(a)]

Who may be the control operator of an amateur station transmitting in RACES to assist relief operations during a disaster?

A. Only a person holding an FCC issued amateur operator license

G2B10 (D) [97.407(b)]

When may the FCC restrict normal frequency operations of amateur stations participating in RACES?

D. When the President's War Emergency Powers have been invoked

G2B11 (A) [97.405]

What frequency should be used to send a distress call?

A. Whatever frequency has the best chance of communicating the distress message

G2B12 (C) [97.405(b)]

When is an amateur station allowed to use any means at its disposal to assist another station in distress?

C. At any time during an actual emergency

G2C - CW operating procedures and procedural signals; Q signals and common abbreviations: full break in

G2C01 (D)

Which of the following describes full break-in telegraphy (QSK)?

D. Transmitting stations can receive between code characters and elements

G2C02 (A)

What should you do if a CW station sends "QRS"?

A. Send slower

G2C03 (C)

What does it mean when a CW operator sends "KN" at the end of a transmission?

C. Listening only for a specific station or stations

G2C04 (D)

What does it mean when a CW operator sends "CL" at the end of a transmission?

D. Closing station

G2C05 (B)

What is the best speed to use answering a CQ in Morse Code?

B. The speed at which the CQ was sent

G2C06 (D)

What does the term "zero beat" mean in CW operation?

D. Matching your transmit frequency to the frequency of a received signal.

G2C07 (A)

When sending CW, what does a "C" mean when added to the RST report?

A. Chirpy or unstable signal

G2C08 (C)

What prosign is sent to indicate the end of a formal message when using CW?

C. AR

G2C09 (C)

What does the Q signal "QSL" mean?

C. I acknowledge receipt

G2C10 (B)

What does the Q signal "QRQ" mean?

B. Send faster

G2C11 (D)

What does the Q signal "QRV" mean?

D. I am ready to receive messages

G2D - Amateur Auxiliary; minimizing interference; HF operations

G2D01 (A)

What is the Amateur Auxiliary to the FCC?

A. Amateur volunteers who are formally enlisted to monitor the airwaves for rules violations

G2D02 (B)

Which of the following are objectives of the Amateur Auxiliary?

B. To encourage amateur self regulation and compliance with the rules

G2D03 (B)

What skills learned during "hidden transmitter hunts" are of help to the Amateur Auxiliary?

B. Direction finding used to locate stations violating FCC Rules

G2D04 (B)

Which of the following describes an azimuthal projection map?

B. A world map projection centered on a particular location

G2D05 (B) [97.111(a)(1)]

When is it permissible to communicate with amateur stations in countries outside the areas administered by the Federal Communications Commission?

B. When the contact is with amateurs in any country except those whose administrations have notified the ITU that they object to such communications

G2D06 (C)

How is a directional antenna pointed when making a "long-path" contact with another station?

C. 180 degrees from its short-path heading

G2D07 (A) [97.303s]

Which of the following is required by the FCC rules when operating in the 60 meter band?

A. If you are using other than a dipole antenna, you must keep a record of the gain of your antenna

G2D08 (D)

Why do many amateurs keep a log even though the FCC doesn't require it?

D. To help with a reply if the FCC requests information

G2D09 (D)

What information is traditionally contained in a station log?

A. Date and time of contact

B. Band and/or frequency of the contact

C. Call sign of station contacted and the signal report given

D. All of these choices are correct

G2D10 (B)

What is QRP operation?

B. Low power transmit operation

G2D11 (C)

Which HF antenna would be the best to use for minimizing interference?

C. A unidirectional antenna

G2E - Digital operating: procedures, procedural signals and common abbreviations

G2E01 (D)

Which mode is normally used when sending an RTTY signal via AFSK with an SSB transmitter?

D. LSB

G2E02 (A)

How many data bits are sent in a single PSK31 character?

A. The number varies

G2E03 (C)

What part of a data packet contains the routing and handling information?

C. Header

G2E04 (B)

What segment of the 20 meter band is most often used for data transmissions?

B. 14.070 - 14.100 MHz

G2E05 (C)

Which of the following describes Baudot code?

C. A 5-bit code with additional start and stop bits

G2E06 (B)

What is the most common frequency shift for RTTY emissions in the amateur HF bands?

B. 170 Hz

G2E07 (B)

What does the abbreviation "RTTY" stand for?

B. Radioteletype

G2E08 (A)

What segment of the 80 meter band is most commonly used for data transmissions?

A. 3570 – 3600 kHz

G2E09 (D)

In what segment of the 20 meter band are most PSK31 operations commonly found?

D. Below the RTTY segment, near 14.070 MHz

G2E10 (D)

What is a major advantage of MFSK16 compared to other digital modes?

D. It offers good performance in weak signal environments without error correction

G2E11 (B)

What does the abbreviation "MFSK" stand for?

B. Multi (or Multiple) Frequency Shift Keying

G2E12 (B)

How does the receiving station respond to an ARQ data mode packet containing errors?

B. Requests the packet be retransmitted

G2E13 (A)

In the PACTOR protocol, what is meant by an NAK response to a transmitted packet?

A. The receiver is requesting the packet be re-transmitted

SUBELEMENT G3 - RADIO WAVE PROPAGATION [3 Exam Questions - 3 Groups]

G3A - Sunspots and solar radiation; ionospheric disturbances; propagation forecasting and indices

G3A01 (A)

What is the sunspot number?

A. A measure of solar activity based on counting sunspots and sunspot groups

G3A02 (B)

What effect does a Sudden Ionospheric Disturbance have on the daytime ionospheric propagation of HF radio waves?

B. It disrupts signals on lower frequencies more than those on higher frequencies

G3A03 (C)

Approximately how long does it take the increased ultraviolet and X-ray radiation from solar flares to affect radio-wave propagation on the Earth?

C. 8 minutes

G3A04 (D)

Which of the following amateur radio HF frequencies are least reliable for long distance communications during periods of low solar activity?

D. 21 MHz and higher

G3A05 (D) [Modified}

What is the solar-flux index?

D. A measure of solar radiation at 10.7 cm

G3A06 (D)

What is a geomagnetic storm?

D. A temporary disturbance in the Earth's magnetosphere

G3A07 (D)

At what point in the solar cycle does the 20 meter band usually support worldwide propagation during daylight hours?

D. At any point in the solar cycle

G3A08 (B)

Which of the following effects can a geomagnetic storm have on radio-wave propagation?

B. Degraded high-latitude HF propagation

G3A09 (C)

What effect do high sunspot numbers have on radio communications?

C. Long-distance communication in the upper HF and lower VHF range is enhanced

G3A10 (C)

What causes HF propagation conditions to vary periodically in a 28-day cycle?

C. The Sun's rotation on its axis

G3A11 (D)

Approximately how long is the typical sunspot cycle?

D. 11 years

G3A12 (B)

What does the K-index indicate?

B. The short term stability of the Earth's magnetic field

G3A13 (C)

What does the A-index indicate?

C. The long term stability of the Earth's geomagnetic field

G3A14 (B)

How are radio communications usually affected by the charged particles that reach the Earth from solar coronal holes?

B. HF communications are disturbed

G3A15 (D)

How long does it take charged particles from coronal mass ejections to affect radio-wave propagation on the Earth?

D. 20 to 40 hours

G3A16 (A)

What is a possible benefit to radio communications resulting from periods of high geomagnetic activity?

A. Aurora that can reflect VHF signals

G3B - Maximum Usable Frequency; Lowest Usable Frequency; propagation

G3B01 (D)

How might a sky-wave signal sound if it arrives at your receiver by both short path and long path propagation?

D. A well-defined echo might be heard

G3B02 (A)

Which of the following is a good indicator of the possibility of sky-wave propagation on the 6 meter band?

A. Short skip sky-wave propagation on the 10 meter band

G3B03 (A)

Which of the following applies when selecting a frequency for lowest attenuation when transmitting on HF?

A. Select a frequency just below the MUF

G3B04 (A)

What is a reliable way to determine if the Maximum Usable Frequency (MUF) is high enough to support skip propagation between your station and a distant location on frequencies between 14 and 30 MHz?

A. Listen for signals from an international beacon

G3B05 (A)

What usually happens to radio waves with frequencies below the Maximum Usable Frequency (MUF) and above the Lowest Usable Frequency (LUF) when they are sent into the ionosphere?

A. They are bent back to the Earth

G3B06 (C)

What usually happens to radio waves with frequencies below the Lowest Usable Frequency (LUF)?

C. They are completely absorbed by the ionosphere

G3B07 (A)

What does LUF stand for?

A. The Lowest Usable Frequency for communications between two points

G3B08 (B)

What does MUF stand for?

B. The Maximum Usable Frequency for communications between two points

G3B09 (C)

What is the approximate maximum distance along the Earth's surface that is normally covered in one hop using the F2 region?

C. 2,500 miles

G3B10 (B)

What is the approximate maximum distance along the Earth's surface that is normally covered in one hop using the E region?

B. 1,200 miles

G3B11 (A)

What happens to HF propagation when the Lowest Usable Frequency (LUF) exceeds the Maximum Usable Frequency (MUF)?

A.   No HF radio frequency will support ordinary skywave communications over the path

G3B12 (D)

What factors affect the Maximum Usable Frequency (MUF)?

A. Path distance and location

B. Time of day and season

C. Solar radiation and ionospheric disturbances

D. All of these choices are correct

G3C - Ionospheric layers; critical angle and frequency; HF scatter; Near Vertical Incidence Sky waves

G3C01 (A)

Which of the following ionospheric layers is closest to the surface of the Earth?

A. The D layer

G3C02 (A)

Where on the Earth do ionospheric layers reach their maximum height?

A. Where the Sun is overhead

G3C03 (C)

Why is the F2 region mainly responsible for the longest distance radio wave propagation?

C. Because it is the highest ionospheric region

G3C04 (D)

What does the term "critical angle" mean as used in radio wave propagation?

D. The highest takeoff angle that will return a radio wave to the Earth under specific ionospheric conditions

G3C05 (C)

Why is long distance communication on the 40, 60, 80 and 160 meter bands more difficult during the day?

C. The D layer absorbs signals at these frequencies during daylight hours

G3C06 (B)

What is a characteristic of HF scatter signals?

B. They have a wavering sound

G3C07 (D)

What makes HF scatter signals often sound distorted?

D. Energy is scattered into the skip zone through several different radio wave paths

G3C08 (A)

Why are HF scatter signals in the skip zone usually weak?

A. Only a small part of the signal energy is scattered into the skip zone

G3C09 (B)

What type of radio wave propagation allows a signal to be detected at a distance too far for ground wave propagation but too near for normal sky-wave propagation?

B. Scatter

G3C10 (D)

Which of the following might be an indication that signals heard on the HF bands are being received via scatter propagation?

D. The signal is heard on a frequency above the Maximum Usable Frequency

G3C11 (B)

Which of the following antenna types will be most effective for skip communications on 40 meters during the day?

B. Horizontal dipoles placed between 1/8 and 1/4 wavelength above the ground

G3C12 (D)

Which ionospheric layer is the most absorbent of long skip signals during daylight hours on frequencies below 10 MHz?

D. The D layer

G3C13 (B)

What is Near Vertical Incidence Sky-wave (NVIS) propagation?

B. Short distance HF propagation using high elevation angles

SUBELEMENT G4 - AMATEUR RADIO PRACTICES [5 Exam Questions - 5 groups]

G4A – Station Operation and set up

G4A01 (B)

What is the purpose of the "notch filter" found on many HF transceivers?

B. To reduce interference from carriers in the receiver passband

G4A02 (C)

What is one advantage of selecting the opposite or "reverse" sideband when receiving CW signals on a typical HF transceiver?

C. It may be possible to reduce or eliminate interference from other signals

G4A03 (C)

What is normally meant by operating a transceiver in "split" mode?

C. The transceiver is set to different transmit and receive frequencies

G4A04 (B)

What reading on the plate current meter of a vacuum tube RF power amplifier indicates correct adjustment of the plate tuning control?

B. A pronounced dip

G4A05 (C)

What is a purpose of using Automatic Level Control (ALC) with a RF power amplifier?

C. To reduce distortion due to excessive drive

G4A06 (C)

What type of device is often used to enable matching the transmitter output to an impedance other than 50 ohms?

C. Antenna coupler

G4A07 (D)

What condition can lead to permanent damage when using a solid-state RF power amplifier?

D. Excessive drive power

G4A08 (D)

What is the correct adjustment for the load or coupling control of a vacuum tube RF power amplifier?

D. Maximum power output without exceeding maximum allowable plate current

G4A09 (C)

Why is a time delay sometimes included in a transmitter keying circuit?

C. To allow time for transmit-receive changeover operations to complete properly before RF output is allowed

G4A10 (B)

What is the purpose of an electronic keyer?

B. Automatic generation of strings of dots and dashes for CW operation

G4A11 (A)

Which of the following is a use for the IF shift control on a receiver?

A. To avoid interference from stations very close to the receive frequency

G4A12 (C)

Which of the following is a common use for the dual VFO feature on a transceiver?

C. To permit ease of monitoring the transmit and receive frequencies when they are not the same

G4A13 (A)

What is one reason to use the attenuator function that is present on many HF transceivers?

A. To reduce signal overload due to strong incoming signals

G4A14 (B)

How should the transceiver audio input be adjusted when transmitting PSK31 data signals?

B. So that the transceiver ALC system does not activate

G4B - Test and monitoring equipment; two-tone test

G4B01 (D)

What item of test equipment contains horizontal and vertical channel amplifiers?

D. An oscilloscope

G4B02 (D)

Which of the following is an advantage of an oscilloscope versus a digital voltmeter?

D. Complex waveforms can be measured

G4B03 (A)

Which of the following is the best instrument to use when checking the keying waveform of a CW transmitter?

A. An oscilloscope

G4B04 (D)

What signal source is connected to the vertical input of an oscilloscope when checking the RF envelope pattern of a transmitted signal?

D. The attenuated RF output of the transmitter

G4B05 (D)

Why is high input impedance desirable for a voltmeter?

B.   It decreases the loading on circuits being measured

G4B06 (C)

What is an advantage of a digital voltmeter as compared to an analog voltmeter?

C. Better precision for most uses

G4B07 (A)

Which of the following might be a use for a field strength meter?

A. Close-in radio direction-finding

G4B08 (A)

Which of the following instruments may be used to monitor relative RF output when making antenna and transmitter adjustments?

A. A field-strength meter

G4B09 (B)

Which of the following can be determined with a field strength meter?

B. The radiation pattern of an antenna

G4B10 (A)

Which of the following can be determined with a directional wattmeter?

A. Standing wave ratio

G4B11 (C)

Which of the following must be connected to an antenna analyzer when it is being used for SWR measurements?

C. Antenna and feed line

G4B12 (B)

What problem can occur when making measurements on an antenna system with an antenna analyzer?

B. Strong signals from nearby transmitters can affect the accuracy of measurements

G4B13 (C)

What is a use for an antenna analyzer other than measuring the SWR of an antenna system?

C. Determining the impedance of an unknown or unmarked coaxial cable

G4B14 (D)

What is an instance in which the use of an instrument with analog readout may be preferred over an instrument with a numerical digital readout?

D. When adjusting tuned circuits

G4B15 (A)

What type of transmitter performance does a two-tone test analyze?

A. Linearity

G4B16 (B)

What signals are used to conduct a two-tone test?

B. Two non-harmonically related audio signals

G4C - Interference with consumer electronics; grounding; DSP

G4C01 (B)

Which of the following might be useful in reducing RF interference to audio-frequency devices?

B. Bypass capacitor

G4C02 (C)

Which of the following could be a cause of interference covering a wide range of frequencies?

C. Arcing at a poor electrical connection

G4C03 (C)

What sound is heard from an audio device or telephone if there is interference from a nearby single-sideband phone transmitter?

C.   Distorted speech

G4C04 (A)

What is the effect on an audio device or telephone system if there is interference from a nearby CW transmitter?

A. On-and-off humming or clicking

G4C05 (D)

What might be the problem if you receive an RF burn when touching your equipment while transmitting on an HF band, assuming the equipment is connected to a ground rod?

D. The ground wire has high impedance on that frequency

G4C06 (C)

What effect can be caused by a resonant ground connection?

C. High RF voltages on the enclosures of station equipment

G4C07 (A)

What is one good way to avoid unwanted effects of stray RF energy in an amateur station?

A. Connect all equipment grounds together

G4C08 (A)

Which of the following would reduce RF interference caused by common-mode current on an audio cable?

A. Placing a ferrite bead around the cable

G4C09 (D)

How can a ground loop be avoided?

D. Connect all ground conductors to a single point

G4C10 (A)

What could be a symptom of a ground loop somewhere in your station?

A. You receive reports of "hum" on your station's transmitted signal

G4C11 (B)

Which of the following is one use for a Digital Signal Processor in an amateur station?

B. To remove noise from received signals

G4C12 (A)

Which of the following is an advantage of a receiver Digital Signal Processor IF filter as compared to an analog filter?

A. A wide range of filter bandwidths and shapes can be created

G4C13 (B)

Which of the following can perform automatic notching of interfering carriers?

B. A Digital Signal Processor (DSP) filter

G4D - Speech processors; S meters; sideband operation near band edges

G4D01 (A)

What is the purpose of a speech processor as used in a modern transceiver?

A. Increase the intelligibility of transmitted phone signals during poor conditions

G4D02 (B)

Which of the following describes how a speech processor affects a transmitted single sideband phone signal?

B. It increases average power

G4D03 (D)

Which of the following can be the result of an incorrectly adjusted speech processor?

A. Distorted speech

B. Splatter

C. Excessive background pickup

D. All of these choices are correct

G4D04 (C)

What does an S meter measure?

C. Received signal strength

G4D05 (D)

How does an S meter reading of 20 dB over S-9 compare to an S-9 signal, assuming a properly calibrated S meter?

D. It is 100 times stronger

G4D06 (A)

Where is an S meter found?

A. In a receiver

G4D07 (C)

How much must the power output of a transmitter be raised to change the S- meter reading on a distant receiver from S8 to S9?

C. Approximately 4 times

G4D08 (C)

What frequency range is occupied by a 3 kHz LSB signal when the displayed carrier frequency is set to 7.178 MHz?

B.   7.175 to 7.178 MHz

G4D09 (B)

What frequency range is occupied by a 3 kHz USB signal with the displayed carrier frequency set to 14.347 MHz?

B. 14.347 to 14.350 MHz

G4D10 (A)

How close to the lower edge of the 40 meter General Class phone segment should your displayed carrier frequency be when using 3 kHz wide LSB?

A. 3 kHz above the edge of the segment

G4D11 (B)

How close to the upper edge of the 20 meter General Class band should your displayed carrier frequency be when using 3 kHz wide USB?

B. 3 kHz below the edge of the band

G4E - HF mobile radio installations; emergency and battery powered operation

G4E01 (C)

What is a "capacitance hat", when referring to a mobile antenna?

C. A device to electrically lengthen a physically short antenna

G4E02 (D)

What is the purpose of a "corona ball" on a HF mobile antenna?

D. To reduce high voltage discharge from the tip of the antenna

G4E03 (A)

Which of the following direct, fused power connections would be the best for a 100-watt HF mobile installation?

A. To the battery using heavy gauge wire

G4E04 (B)

Why is it best NOT to draw the DC power for a 100-watt HF transceiver from an automobile's auxiliary power socket?

B. The socket's wiring may be inadequate for the current being drawn by the transceiver

G4E05 (C)

Which of the following most limits the effectiveness of an HF mobile transceiver operating in the 75 meter band?

C. The antenna system

G4E06 (C)

What is one disadvantage of using a shortened mobile antenna as opposed to a full size antenna?

C. Operating bandwidth may be very limited

G4E07 (D)

Which of the following is the most likely to cause interfering signals to be heard in the receiver of an HF mobile installation in a recent model vehicle?

D. The vehicle control computer

G4E08 (A)

What is the name of the process by which sunlight is changed directly into electricity?

A. Photovoltaic conversion

G4E09 (B)

What is the approximate open-circuit voltage from a modern, well-illuminated photovoltaic cell

C. 0.5 VDC

G4E10 (B)

What is the reason a series diode is connected between a solar panel and a storage battery that is being charged by the panel?

B. The diode prevents self discharge of the battery though the panel during times of low or no illumination

G4E11 (C)

Which of the following is a disadvantage of using wind as the primary source of power for an emergency station?

C. A large energy storage system is needed to supply power when the wind is not blowing

SUBELEMENT G5 – ELECTRICAL PRINCIPLES [3 Exam Questions – 3 Groups]

G5A - Reactance; inductance; capacitance; impedance; impedance matching

G5A01 (C)

What is impedance?

C. The opposition to the flow of current in an AC circuit

G5A02 (B)

What is reactance?

B. Opposition to the flow of alternating current caused by capacitance or inductance

G5A03 (D)

Which of the following causes opposition to the flow of alternating current in an inductor?

D. Reactance

G5A04 (C)

Which of the following causes opposition to the flow of alternating current in a capacitor?

C. Reactance

G5A05 (D)

How does an inductor react to AC?

D. As the frequency of the applied AC increases, the reactance increases

G5A06 (A)

How does a capacitor react to AC?

A. As the frequency of the applied AC increases, the reactance decreases

G5A07 (D)

What happens when the impedance of an electrical load is equal to the internal impedance of the power source?

D. The source can deliver maximum power to the load

G5A08 (A)

Why is impedance matching important?

A. So the source can deliver maximum power to the load

G5A09 (B)

What unit is used to measure reactance?

B. Ohm

G5A10 (B)

What unit is used to measure impedance?

B. Ohm

G5A11 (A)

Which of the following describes one method of impedance matching between two AC circuits?

A. Insert an LC network between the two circuits

G5A12 (B)

What is one reason to use an impedance matching transformer?

B. To maximize the transfer of power

G5A13 (D)

Which of the following devices can be used for impedance matching at radio frequencies?

A. A transformer

B. A Pi-network

C. A length of transmission line

D. All of these choices are correct

G5B - The Decibel; current and voltage dividers; electrical power calculations; sine wave root-mean-square (RMS) values; PEP calculations

G5B01 (B)

A two-times increase or decrease in power results in a change of how many dB?

B. Approximately 3 dB

G5B02 (C)

How does the total current relate to the individual currents in each branch of a parallel circuit?

C. It equals the sum of the currents through each branch

G5B03 (B)

How many watts of electrical power are used if 400 VDC is supplied to an 800-ohm load?

B. 200 watts

G5B04 (A)

How many watts of electrical power are used by a 12-VDC light bulb that draws 0.2 amperes?

A.   2.4 watts

G5B05 (A)

How many watts are dissipated when a current of 7.0 milliamperes flows through 1.25 kilohms?

A. Approximately 61 milliwatts

G5B06 (B)

What is the output PEP from a transmitter if an oscilloscope measures 200 volts peak-to-peak across a 50-ohm dummy load connected to the transmitter output?

B. 100 watts

G5B07 (C)

Which value of an AC signal results in the same power dissipation as a DC voltage of the same value?

C. The RMS value

G5B08 (D)

What is the peak-to-peak voltage of a sine wave that has an RMS voltage of 120 volts?

D. 339.4 volts

G5B09 (B)

What is the RMS voltage of a sine wave with a value of 17 volts peak?

B. 12 volts

G5B10 (C)

What percentage of power loss would result from a transmission line loss of 1 dB?

C. 20.5%

G5B11 (B)

What is the ratio of peak envelope power to average power for an unmodulated carrier?

B.   1.00

G5B12 (B)

What would be the RMS voltage across a 50-ohm dummy load dissipating 1200 watts?

B. 245 volts

G5B13 (B)

What is the output PEP of an unmodulated carrier if an average reading wattmeter connected to the transmitter output indicates 1060 watts?

B. 1060 watts

G5B14 (B)

What is the output PEP from a transmitter if an oscilloscope measures 500 volts peak-to-peak across a 50-ohm resistor connected to the transmitter output?

B. 625 watts

G5C – Resistors, capacitors, and inductors in series and parallel; transformers

G5C01 (C)

What causes a voltage to appear across the secondary winding of a transformer when an AC voltage source is connected across its primary winding?

C. Mutual inductance

G5C02 (B)

Which part of a transformer is normally connected to the incoming source of energy?

B. The primary

G5C03 (B)

Which of the following components should be added to an existing resistor to increase the resistance?

B. A resistor in series

G5C04 (C)

What is the total resistance of three 100-ohm resistors in parallel?

C. 33.3 ohms

G5C05 (C)

If three equal value resistors in parallel produce 50 ohms of resistance, and the same three resistors in series produce 450 ohms, what is the value of each resistor?

C. 150 ohms

G5C06 (C)

What is the RMS voltage across a 500-turn secondary winding in a transformer if the 2250-turn primary is connected to 120 VAC?

C. 26.7 volts

G5C07 (A)

What is the turns ratio of a transformer used to match an audio amplifier having a 600-ohm output impedance to a speaker having a 4-ohm impedance?

A. 12.2 to 1

G5C08 (D)

What is the equivalent capacitance of two 5000 picofarad capacitors and one 750 picofarad capacitor connected in parallel?

D. 10750 picofarads

G5C09 (C)

What is the capacitance of three 100 microfarad capacitors connected in series?

C. 33.3 microfarads

G5C10 (C)

What is the inductance of three 10 millihenry inductors connected in parallel?

C. 3.3 millihenrys

G5C11 (C)

What is the inductance of a 20 millihenry inductor in series with a 50 millihenry inductor?

C. 70 millihenrys

G5C12 (B)

What is the capacitance of a 20 microfarad capacitor in series with a 50 microfarad capacitor?

B. 14.3 microfarads

G5C13 (C)

Which of the following components should be added to a capacitor to increase the capacitance?

C. A capacitor in parallel

G5C14 (D)

Which of the following components should be added to an inductor to increase the inductance?

D. An inductor in series

G5C15 (A)

What is the total resistance of a 10 ohm, a 20 ohm, and a 50 ohm resistor in parallel?

A. 5.9 ohms

SUBELEMENT G6 – CIRCUIT COMPONENTS [3 Exam Questions – 3 Groups]

G6A - Resistors; capacitors; inductors

G6A01 (A)

Which of the following is an important characteristic for capacitors used to filter the DC output of a switching power supply?

A. Low equivalent series resistance

G6A02 (D)

Which of the following types of capacitors are often used in power supply circuits to filter the rectified AC?

D. Electrolytic

G6A03 (D)

Which of the following is an advantage of ceramic capacitors as compared to other types of capacitors?

D. Comparatively low cost

G6A04 (C)

Which of the following is an advantage of an electrolytic capacitor?

C. High capacitance for given volume

G6A05 (A)

Which of the following is one effect of lead inductance in a capacitor used at VHF and above?

A. Effective capacitance may be reduced

G6A06 (C)

What will happen to the resistance if the temperature of a resistor is increased?

C. It will change depending on the resistor's temperature coefficient

G6A07 (B)

Which of the following is a reason not to use wire-wound resistors in an RF circuit?

B. The resistor's inductance could make circuit performance unpredictable

G6A08 (B)

Which of the following describes a thermistor?

B. A device having a specific change in resistance with temperature variations

G6A09 (D)

What is an advantage of using a ferrite core toroidal inductor?

A. Large values of inductance may be obtained

B. The magnetic properties of the core may be optimized for a specific range of frequencies

C. Most of the magnetic field is contained in the core

D. All of these choices are correct

G6A10 (C)

How should the winding axes of solenoid inductors be placed to minimize their mutual inductance?

C. At right angles

G6A11 (B)

Why would it be important to minimize the mutual inductance between two inductors?

B. To reduce unwanted coupling between circuits

G6A12 (D)

What is a common name for an inductor used to help smooth the DC output from the rectifier in a conventional power supply?

D. Filter choke

G6A13 (B)

What is an effect of inter-turn capacitance in an inductor?

B. The inductor may become self resonant at some frequencies

G6B - Rectifiers; solid state diodes and transistors; vacuum tubes; batteries

G6B01 (C)

What is the peak-inverse-voltage rating of a rectifier?

C. The maximum voltage the rectifier will handle in the non-conducting direction

G6B02 (A)

What are two major ratings that must not be exceeded for silicon diode rectifiers?

A. Peak inverse voltage; average forward current

G6B03 (B)

What is the approximate junction threshold voltage of a germanium diode?

B. 0.3 volts

G6B04 (C)

When two or more diodes are connected in parallel to increase current handling capacity, what is the purpose of the resistor connected in series with each diode?

C. To ensure that one diode doesn't carry most of the current

G6B05 (C)

What is the approximate junction threshold voltage of a conventional silicon diode?

C. 0.7 volts

G6B06 (A)

Which of the following is an advantage of using a Schottky diode in an RF switching circuit as compared to a standard silicon diode?

A. Lower capacitance

G6B07 (A)

What are the stable operating points for a bipolar transistor used as a switch in a logic circuit?

A. Its saturation and cut-off regions

G6B08 (D)

Why must the cases of some large power transistors be insulated from ground?

D. To avoid shorting the collector or drain voltage to ground

G6B09 (B)

Which of the following describes the construction of a MOSFET?

B. The gate is separated from the channel with a thin insulating layer

G6B10 (A)

Which element of a triode vacuum tube is used to regulate the flow of electrons between cathode and plate?

A. Control grid

G6B11 (B)

Which of the following solid state devices is most like a vacuum tube in its general operating characteristics?

B. A Field Effect Transistor

G6B12 (A)

What is the primary purpose of a screen grid in a vacuum tube?

A. To reduce grid-to-plate capacitance

G6B13 (B)

What is an advantage of the low internal resistance of nickel-cadmium batteries?

D. High discharge current

G6B14 (C)

What is the minimum allowable discharge voltage for maximum life of a standard 12 volt lead acid battery?

C. 10.5 volts

G6B15 (D)

When is it acceptable to recharge a carbon-zinc primary cell?

D. Never

G6C - Analog and digital integrated circuits (IC's); microprocessors; memory; I/O devices; microwave IC's (MMIC's); display devices

G6C01 (D)

Which of the following is an analog integrated circuit?

D. Linear voltage regulator

G6C02 (B)

What is meant by the term MMIC?

B. Monolithic Microwave Integrated Circuit

G6C03 (A)

Which of the following is an advantage of CMOS integrated circuits compared to TTL integrated circuits?

A. Low power consumption

G6C04 (B)

What is meant by the term ROM?

B. Read Only Memory

G6C05 (C)

What is meant when memory is characterized as "non-volatile"?

C. The stored information is maintained even if power is removed

G6C06 (D)

Which of the following describes an integrated circuit operational amplifier?

D. Analog

G6C07 (D)

What is one disadvantage of an incandescent indicator compared to an LED?

D. High power consumption

G6C08 (D)

How is an LED biased when emitting light?

D. Forward Biased

G6C09 (A)

Which of the following is a characteristic of a liquid crystal display?

A. It requires ambient or back lighting

G6C10 (A)

What two devices in an Amateur Radio station might be connected using a USB interface?

A. Computer and transceiver

G6C11 (B)

What is a microprocessor?

B. A computer on a single integrated circuit

G6C12 (D)

Which of the following connectors would be a good choice for a serial data port?

D. DE-9

G6C13 (C)

Which of these connector types is commonly used for RF service at frequencies up to 150 MHz?

C. PL-259

G6C14 (C)

Which of these connector types is commonly used for audio signals in Amateur Radio stations?

C. RCA Phono

G6C15 (B)

What is the main reason to use keyed connectors instead of non-keyed types?

B. Reduced chance of incorrect mating

G6C16 (A)

Which of the following describes a type-N connector?

A. A moisture-resistant RF connector useful to 10 GHz

G6C17 (C)

What is the general description of a DIN type connector?

C. A family of multiple circuit connectors suitable for audio and control signals

G6C18 (B)

What is a type SMA connector?

B. A small threaded connector suitable for signals up to several GHz

G7 – PRACTICAL CIRCUITS [3 Exam Questions – 3 Groups]

G7A Power supplies; and schematic symbols

G7A01 (B)

What safety feature does a power-supply bleeder resistor provide?

B. It discharges the filter capacitors

G7A02 (D)

Which of the following components are used in a power-supply filter network?

D. Capacitors and inductors

G7A03 (D)

What is the peak-inverse-voltage across the rectifiers in a full-wave bridge power supply?

D. Equal to the normal peak output voltage of the power supply

G7A04 (D)

What is the peak-inverse-voltage across the rectifier in a half-wave power supply?

D. Two times the normal peak output voltage of the power supply

G7A05 (B)

What portion of the AC cycle is converted to DC by a half-wave rectifier?

B. 180 degrees

G7A06 (D)

What portion of the AC cycle is converted to DC by a full-wave rectifier?

D. 360 degrees

G7A07 (A)

What is the output waveform of an unfiltered full-wave rectifier connected to a resistive load?

A. A series of DC pulses at twice the frequency of the AC input

G7A08 (C)

Which of the following is an advantage of a switch-mode power supply as compared to a linear power supply?

C. High frequency operation allows the use of smaller components

G7A09 (C)

Which symbol in figure G7-1 represents a field effect transistor?

C. Symbol 1

G7A10 (D)

Which symbol in figure G7-1 represents a Zener diode?

D. Symbol 5

G7A11 (B)

Which symbol in figure G7-1 represents an NPN junction transistor?

B. Symbol 2

G7A12 (C)

Which symbol in Figure G7-1 represents a multiple-winding transformer?

C. Symbol 6

G7A13 (A)

Which symbol in Figure G7-1 represents a tapped inductor?

A. Symbol 7

G7B - Digital circuits; amplifiers and oscillators

G7B01 (A)

Complex digital circuitry can often be replaced by what type of integrated circuit?

A. Microcontroller

G7B02 (A)

Which of the following is an advantage of using the binary system when processing digital signals?

A. Binary "ones" and "zeros" are easy to represent with an "on" or "off" state

G7B03 (B)

Which of the following describes the function of a two input AND gate?

B. Output is high only when both inputs are high

G7B04 (C)

Which of the following describes the function of a two input NOR gate?

C. Output is low when either or both inputs are high

G7B05 (C)

How many states does a 3-bit binary counter have?

C. 8

G7B06 (A)

What is a shift register?

A. A clocked array of circuits that passes data in steps along the array

G7B07 (D)

What are the basic components of virtually all sine wave oscillators?

D. A filter and an amplifier operating in a feedback loop

G7B08 (B)

How is the efficiency of an RF power amplifier determined?

B. Divide the RF output power by the DC input power

G7B09 (C)

What determines the frequency of an LC oscillator?

C. The inductance and capacitance in the tank circuit

G7B10 (D)

Which of the following is a characteristic of a Class A amplifier?

D. Low distortion

G7B11 (B)

For which of the following modes is a Class C power stage appropriate for amplifying a modulated signal?

B. CW

G7B12 (D)

Which of these classes of amplifiers has the highest efficiency?

D. Class C

G7B13 (B)

What is the reason for neutralizing the final amplifier stage of a transmitter?

B. To eliminate self-oscillations

G7B14 (B)

Which of the following describes a linear amplifier?

B. An amplifier in which the output preserves the input waveform

G7C - Receivers and transmitters; filters, oscillators

G7C01 (B)

Which of the following is used to process signals from the balanced modulator and send them to the mixer in a single-sideband phone transmitter?

B. Filter

G7C02 (D)

Which circuit is used to combine signals from the carrier oscillator and speech amplifier and send the result to the filter in a typical single-sideband phone transmitter?

D. Balanced modulator

G7C03 (C)

What circuit is used to process signals from the RF amplifier and local oscillator and send the result to the IF filter in a superheterodyne receiver?

C. Mixer

G7C04 (D)

What circuit is used to combine signals from the IF amplifier and BFO and send the result to the AF amplifier in a single-sideband receiver?

D. Product detector

G7C05 (D)

Which of the following is an advantage of a transceiver controlled by a direct digital synthesizer (DDS)?

D. Variable frequency with the stability of a crystal oscillator

G7C06 (B)

What should be the impedance of a low-pass filter as compared to the impedance of the transmission line into which it is inserted?

B. About the same

G7C07 (C)

What is the simplest combination of stages that implement a superheterodyne receiver?

C. HF oscillator, mixer, detector

G7C08 (D)

What type of circuit is used in many FM receivers to convert signals coming from the IF amplifier to audio?

D. Discriminator

G7C09 (D)

Which of the following is needed for a Digital Signal Processor IF filter?

A. An analog to digital converter

B. A digital to analog converter

C. A digital processor chip

D. All of the these choices are correct

G7C10 (B)

How is Digital Signal Processor filtering accomplished?

B. By converting the signal from analog to digital and using digital processing

G7C11 (A)

What is meant by the term "software defined radio" (SDR)?

A. A radio in which most major signal processing functions are performed by software

SUBELEMENT G8 – SIGNALS AND EMISSIONS [2 Exam Questions – 2 Groups]

G8A - Carriers and modulation: AM; FM; single and double sideband; modulation envelope; overmodulation

G8A01 (D)

What is the name of the process that changes the envelope of an RF wave to carry information?

D. Amplitude modulation

G8A02 (B)

What is the name of the process that changes the phase angle of an RF wave to convey information?

B. Phase modulation

G8A03 (D)

What is the name of the process which changes the frequency of an RF wave to convey information?

D. Frequency modulation

G8A04 (B)

What emission is produced by a reactance modulator connected to an RF power amplifier?

B. Phase modulation

G8A05 (D)

What type of modulation varies the instantaneous power level of the RF signal?

D. Amplitude modulation

G8A06 (C)

What is one advantage of carrier suppression in a single-sideband phone transmission?

C. The available transmitter power can be used more effectively

G8A07 (A)

Which of the following phone emissions uses the narrowest frequency bandwidth?

A. Single sideband

G8A08 (D)

Which of the following is an effect of over-modulation?

D. Excessive bandwidth

G8A09 (B)

What control is typically adjusted for proper ALC setting on an amateur single sideband transceiver?

B. Transmit audio or microphone gain

G8A10 (C)

What is meant by flat-topping of a single-sideband phone transmission?

B.   Signal distortion caused by excessive drive

G8A11 (A)

What happens to the RF carrier signal when a modulating audio signal is applied to an FM transmitter?

A. The carrier frequency changes proportionally to the instantaneous amplitude of the modulating signal

G8A12 (A)

What signal(s) would be found at the output of a properly adjusted balanced modulator?

A. Both upper and lower sidebands

G8B - Frequency mixing; multiplication; HF data communications; bandwidths of various modes; deviation

G8B01 (A)

What receiver stage combines a 14.250 MHz input signal with a 13.795 MHz oscillator signal to produce a 455 kHz intermediate frequency (IF) signal?

A. Mixer

G8B02 (B)

If a receiver mixes a 13.800 MHz VFO with a 14.255 MHz received signal to produce a 455 kHz intermediate frequency (IF) signal, what type of interference will a 13.345 MHz signal produce in the receiver?

B. Image response

G8B03 (A)

What is another term for the mixing of two RF signals?

A. Heterodyning

G8B04 (D)

What is the name of the stage in a VHF FM transmitter that generates a harmonic of a lower frequency signal to reach the desired operating frequency?

D. Multiplier

G8B05 (C)

Why isn't frequency modulated (FM) phone used below 29.5 MHz?

C. The wide bandwidth is prohibited by FCC rules

G8B06 (D)

What is the total bandwidth of an FM-phone transmission having a 5 kHz deviation and a 3 kHz modulating frequency?

D. 16 kHz

G8B07 (B)

What is the frequency deviation for a 12.21-MHz reactance-modulated oscillator in a 5-kHz deviation, 146.52-MHz FM-phone transmitter?

B. 416.7 Hz

G8B08 (B)

Why is it important to know the duty cycle of the data mode you are using when transmitting?

B. Some modes have high duty cycles which could exceed the transmitter's average power rating.

G8B09 (D)

Why is it good to match receiver bandwidth to the bandwidth of the operating mode?

D. It results in the best signal to noise ratio

G8B10 (A)

What does the number 31 represent in PSK31?

A. The approximate transmitted symbol rate

G8B11 (C)

How does forward error correction allow the receiver to correct errors in received data packets?

C. By transmitting redundant information with the data

G8B12 (B)

What is the relationship between transmitted symbol rate and bandwidth?

B. Higher symbol rates require higher bandwidth

SUBELEMENT G9 – ANTENNAS AND FEED LINES [4 Exam Questions – 4 Groups]

G9A - Antenna feed lines: characteristic impedance, and attenuation; SWR calculation, measurement and effects; matching networks

G9A01 (A)

Which of the following factors determine the characteristic impedance of a parallel conductor antenna feed line?

A. The distance between the centers of the conductors and the radius of the conductors

G9A02 (B)

What are the typical characteristic impedances of coaxial cables used for antenna feed lines at amateur stations?

B. 50 and 75 ohms

G9A03 (D)

What is the characteristic impedance of flat ribbon TV type twinlead?

D. 300 ohms

G9A04 (C)

What is the reason for the occurrence of reflected power at the point where a feed line connects to an antenna?

C. A difference between feed-line impedance and antenna feed-point impedance

G9A05 (B)

How does the attenuation of coaxial cable change as the frequency of the signal it is carrying increases?

B. It increases

G9A06 (D)

In what values are RF feed line losses usually expressed?

D. dB per 100 ft

G9A07 (D)

What must be done to prevent standing waves on an antenna feed line?

D. The antenna feed-point impedance must be matched to the characteristic impedance of the feed line

G9A08 (B)

If the SWR on an antenna feed line is 5 to 1, and a matching network at the transmitter end of the feed line is adjusted to 1 to 1 SWR, what is the resulting SWR on the feed line?

B. 5 to 1

G9A09 (A)

What standing wave ratio will result from the connection of a 50-ohm feed line to a non-reactive load having a 200-ohm impedance?

A. 4:1

G9A10 (D)

What standing wave ratio will result from the connection of a 50-ohm feed line to a non-reactive load having a 10-ohm impedance?

D. 5:1

G9A11 (B)

What standing wave ratio will result from the connection of a 50-ohm feed line to a non-reactive load having a 50-ohm impedance?

B. 1:1

G9A12 (A)

What would be the SWR if you feed a vertical antenna that has a 25-ohm feed-point impedance with 50-ohm coaxial cable?

A. 2:1

G9A13 (C)

What would be the SWR if you feed an antenna that has a 300-ohm feed-point impedance with 50-ohm coaxial cable?

C. 6:1

G9B - Basic antennas

G9B01 (B)

What is one disadvantage of a directly fed random-wire antenna?

B. You may experience RF burns when touching metal objects in your station

G9B02 (D)

What is an advantage of downward sloping radials on a quarter wave ground-plane antenna?

D. They bring the feed-point impedance closer to 50 ohms

G9B03 (B)

What happens to the feed-point impedance of a ground-plane antenna when its radials are changed from horizontal to downward-sloping?

B. It increases

G9B04 (A)

What is the low angle azimuthal radiation pattern of an ideal half-wavelength dipole antenna installed 1/2 wavelength high and parallel to the Earth?

A. It is a figure-eight at right angles to the antenna

G9B05 (C)

How does antenna height affect the horizontal (azimuthal) radiation pattern of a horizontal dipole HF antenna?

C. If the antenna is less than 1/2 wavelength high, the azimuthal pattern is almost omnidirectional

G9B06 (C)

Where should the radial wires of a ground-mounted vertical antenna system be placed?

C. On the surface or buried a few inches below the ground

G9B07 (B)

How does the feed-point impedance of a 1/2 wave dipole antenna change as the antenna is lowered from 1/4 wave above ground?

B. It steadily decreases

G9B08 (A)

How does the feed-point impedance of a 1/2 wave dipole change as the feed-point location is moved from the center toward the ends?

A. It steadily increases

G9B09 (A)

Which of the following is an advantage of a horizontally polarized as compared to vertically polarized HF antenna?

A. Lower ground reflection losses

G9B10 (D)

What is the approximate length for a 1/2-wave dipole antenna cut for 14.250 MHz?

D. 32 feet

G9B11 (C)

What is the approximate length for a 1/2-wave dipole antenna cut for 3.550 MHz?

C. 131 feet

G9B12 (A)

What is the approximate length for a 1/4-wave vertical antenna cut for 28.5 MHz?

A. 8 feet

G9C - Directional antennas

G9C01 (A)

Which of the following would increase the bandwidth of a Yagi antenna?

A. Larger diameter elements

G9C02 (B)

What is the approximate length of the driven element of a Yagi antenna?

B. 1/2 wavelength

G9C03 (B)

Which statement about a three-element, single-band Yagi antenna is true?

B. The director is normally the shortest parasitic element

G9C04 (A)

Which statement about a three-element; single-band Yagi antenna is true?

A. The reflector is normally the longest parasitic element

G9C05 (A)

How does increasing boom length and adding directors affect a Yagi antenna?

A. Gain increases

G9C06 (C)

Which of the following is a reason why a Yagi antenna is often used for radio communications on the 20 meter band?

C. It helps reduce interference from other stations to the side or behind the antenna

G9C07 (C)

What does "front-to-back ratio" mean in reference to a Yagi antenna?

C. The power radiated in the major radiation lobe compared to the power radiated in exactly the opposite direction

G9C08 (D)

What is meant by the "main lobe" of a directive antenna?

D. The direction of maximum radiated field strength from the antenna

G9C09 (A)

What is the approximate maximum theoretical forward gain of a three element, single-band Yagi antenna?

A. 9.7 dBi

G9C10 (D)

Which of the following is a Yagi antenna design variable that could be adjusted to optimize forward gain, front-to-back ratio, or SWR bandwidth?

A. The physical length of the boom

B. The number of elements on the boom

C. The spacing of each element along the boom

D. All of these choices are correct

G9C11 (A)

What is the purpose of a gamma match used with Yagi antennas?

A. To match the relatively low feed-point impedance to 50 ohms

G9C12 (A)

Which of the following is an advantage of using a gamma match for impedance matching of a Yagi antenna to 50-ohm coax feed line?

A. It does not require that the elements be insulated from the boom

G9C13 (A)

Approximately how long is each side of a quad antenna driven element?

A. 1/4 wavelength

G9C14 (B)

How does the forward gain of a two-element quad antenna compare to the forward gain of a three-element Yagi antenna?

B. About the same

G9C15 (B)

Approximately how long is each side of a quad antenna reflector element?

B. Slightly more than 1/4 wavelength

G9C16 (D)

How does the gain of a two-element delta-loop beam compare to the gain of a two-element quad antenna?

D. About the same

G9C17 (B)

Approximately how long is each leg of a symmetrical delta-loop antenna?

B. 1/3 wavelength

G9C18 (A)

What happens when the feed point of a quad antenna is changed from the center of either horizontal wire to the center of either vertical wire?

A. The polarization of the radiated signal changes from horizontal to vertical

G9C19 (D)

What configuration of the loops of a two-element quad antenna must be used for the antenna to operate as a beam antenna, assuming one of the elements is used as a reflector?

D. The reflector element must be approximately 5% longer than the driven element

G9C20 (B)

How does the gain of two 3-element horizontally polarized Yagi antennas spaced vertically 1/2 wavelength apart typically compare to the gain of a single 3-element Yagi?

B. Approximately 3 dB higher

G9D - Specialized antennas

G9D01 (D)

What does the term "NVIS" mean as related to antennas?

D. Near Vertical Incidence Sky wave

G9D02 (B)

Which of the following is an advantage of an NVIS antenna?

B. High vertical angle radiation for working stations within a radius of a few hundred kilometers

G9D03 (D)

At what height above ground is an NVIS antenna typically installed?

D. Between 1/10 and 1/4 wavelength

G9D04 (A)

What is the primary purpose of antenna traps?

A. To permit multiband operation

G9D05 (D)

What is the advantage of vertical stacking of horizontally polarized Yagi antennas?

D. Narrows the main lobe in elevation

G9D06 (A)

Which of the following is an advantage of a log periodic antenna?

A. Wide bandwidth

G9D07 (A)

Which of the following describes a log periodic antenna?

A. Length and spacing of the elements increases logarithmically from one end of the boom to the other

G9D08 (B)

Why is a Beverage antenna not used for transmitting?

B. It has high losses compared to other types of antennas

G9D09 (B)

Which of the following is an application for a Beverage antenna?

B. Directional receiving for low HF bands

G9D10 (D)

Which of the following describes a Beverage antenna?

D. A very long and low directional receiving antenna

G9D11 (D)

Which of the following is a disadvantage of multiband antennas?

D. They have poor harmonic rejection

SUBELEMENT G0 – ELECTRICAL AND RF SAFETY [2 Exam Questions – 2 Groups]

G0A - RF safety principles, rules and guidelines; routine station evaluation

G0A01 (A)

What is one way that RF energy can affect human body tissue?

A. It heats body tissue

G0A02 (D)

Which of the following properties is important in estimating whether an RF signal exceeds the maximum permissible exposure (MPE)?

A. Its duty cycle

B. Its frequency

C. Its power density

D. All of these choices are correct

G0A03 (D) [97.13(c)(1)]

How can you determine that your station complies with FCC RF exposure regulations?

A. By calculation based on FCC OET Bulletin 65

B. By calculation based on computer modeling

C. By measurement of field strength using calibrated equipment

D. All of these choices are correct

G0A04 (D)

What does "time averaging" mean in reference to RF radiation exposure?

D. The total RF exposure averaged over a certain time

G0A05 (A)

What must you do if an evaluation of your station shows RF energy radiated from your station exceeds permissible limits?

A. Take action to prevent human exposure to the excessive RF fields

G0A06 (C)—Question withdrawn by QPC 2/22/2011

Which transmitter(s) at a multiple user site is/are responsible for RF safety compliance?

C. Any transmitter that contributes 5% or more of the MPE

G0A07 (A)

What effect does transmitter duty cycle have when evaluating RF exposure?

A. A lower transmitter duty cycle permits greater short-term exposure levels

G0A08 (C)

Which of the following steps must an amateur operator take to ensure compliance with RF safety regulations when transmitter power exceeds levels specified in part 97.13?

C. Perform a routine RF exposure evaluation

G0A09 (B)

What type of instrument can be used to accurately measure an RF field?

B. A calibrated field-strength meter with a calibrated antenna

G0A10 (D)

What is one thing that can be done if evaluation shows that a neighbor might receive more than the allowable limit of RF exposure from the main lobe of a directional antenna?

D. Take precautions to ensure that the antenna cannot be pointed in their direction

G0A11 (C)

What precaution should you take if you install an indoor transmitting antenna?

C. Make sure that MPE limits are not exceeded in occupied areas

G0A12 (B)

What precaution should you take whenever you make adjustments or repairs to an antenna?

B. Turn off the transmitter and disconnect the feed line

G0A13 (D)

What precaution should be taken when installing a ground-mounted antenna?

D. It should be installed so no one can be exposed to RF radiation in excess of maximum permissible limits

G0B - Safety in the ham shack: electrical shock and treatment, safety grounding, fusing, interlocks, wiring, antenna and tower safety

G0B01 (A)

Which wire or wires in a four-conductor line cord should be attached to fuses or circuit breakers in a device operated from a 240-VAC single-phase source?

A. Only the hot wires

G0B02 (C)

What is the minimum wire size that may be safely used for a circuit that draws up to 20 amperes of continuous current?

C. AWG number 12

G0B03 (D)

Which size of fuse or circuit breaker would be appropriate to use with a circuit that uses AWG number 14 wiring?

D. 15 amperes

G0B04 (A)

Which of the following is a primary reason for not placing a gasoline-fueled generator inside an occupied area?

A. Danger of carbon monoxide poisoning

G0B05 (B)

Which of the following conditions will cause a Ground Fault Circuit Interrupter (GFCI) to disconnect the 120 or 240 Volt AC line power to a device?

B. Current flowing from one or more of the hot wires directly to ground

G0B06 (D)

Why must the metal enclosure of every item of station equipment be grounded?

D. It ensures that hazardous voltages cannot appear on the chassis

G0B07 (B)

Which of the following should be observed for safety when climbing on a tower using a safety belt or harness?

B. Always attach the belt safety hook to the belt D-ring with the hook opening away from the tower

G0B08 (B)

What should be done by any person preparing to climb a tower that supports electrically powered devices?

B. Make sure all circuits that supply power to the tower are locked out and tagged

G0B09 (D)

Why should soldered joints not be used with the wires that connect the base of a tower to a system of ground rods?

D. A soldered joint will likely be destroyed by the heat of a lightning strike

G0B10 (A)

Which of the following is a danger from lead-tin solder?

A. Lead can contaminate food if hands are not washed carefully after handling

G0B11 (D)

Which of the following is good engineering practice for lightning protection grounds?

D. They must be bonded together with all other grounds

G0B12 (C)

What is the purpose of a transmitter power supply interlock?

C. To ensure that dangerous voltages are removed if the cabinet is opened

G0B13 (A)

What must you do when powering your house from an emergency generator?

A. Disconnect the incoming utility power feed

G0B14 (C)

Which of the following is covered by the National Electrical Code?

C. Electrical safety inside the ham shack

G0B15 (A)

Which of the following is true of an emergency generator installation?

A. The generator should be located in a well ventilated area

G0B16 (C)

When might a lead acid storage battery give off explosive hydrogen gas?

C. When being charged

2011-2015 Element 3 Figure G7-1

Revised 2-22-2011 Figure G7-1 was replaced to correct schematic error in which the varactor diode (item 4) was shown reversed. (See new G7-1 files on the website: www.ncvec.org for additional formats available for download.)

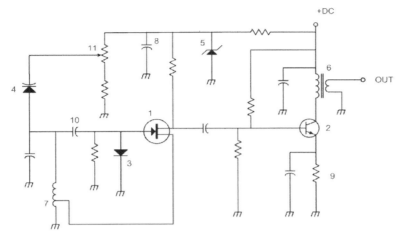

Figure G7-1

~~ End of Pool ~~

The Complete Study Guide For All Amateur Radio Tests

EXTRA CLASS

REVISED January 31, 2012

2012-2016 ELEMENT 4 EXTRA CLASS QUESTION POOL

Effective July 1, 2012 to June 30, 2016

EXTRA CLASS QUESTION POOL

The Complete Study Guide For All Amateur Radio Tests

SUBELEMENT E1 - COMMISSION'S RULES [6 Exam Questions - 6 Groups]

E1A Operating Standards: frequency privileges; emission standards; automatic message forwarding; frequency sharing; stations aboard ships or aircraft

E1A01 (D) [97.301, 97.305]

When using a transceiver that displays the carrier frequency of phone signals, which of the following displayed frequencies represents the highest frequency at which a properly adjusted USB emission will be totally within the band?

D. 3 kHz below the upper band edge

E1A02 (D) [97.301, 97.305]

When using a transceiver that displays the carrier frequency of phone signals, which of the following displayed frequencies represents the lowest frequency at which a properly adjusted LSB emission will be totally within the band?

C.   3 kHz above the lower band edge

E1A03 (C) [97.301, 97.305]

With your transceiver displaying the carrier frequency of phone signals, you hear a DX station's CQ on 14.349 MHz USB. Is it legal to return the call using upper sideband on the same frequency?

C. No, my sidebands will extend beyond the band edge

E1A04 (C) [97.301, 97.305]

With your transceiver displaying the carrier frequency of phone signals, you hear a DX station calling CQ on 3.601 MHz LSB. Is it legal to return the call using lower sideband on the same frequency?

C. No, my sidebands will extend beyond the edge of the phone band segment

E1A05 (C) [97.313]

What is the maximum power output permitted on the 60 meter band?

C. 100 watts PEP effective radiated power relative to the gain of a half-wave dipole

E1A06 (B) [97.303]

Which of the following describes the rules for operation on the 60 meter band?

B. Operation is restricted to specific emission types and specific channels

E1A07 (D) [97.303]

What is the only amateur band where transmission on specific channels rather than a range of frequencies is permitted?

D. 60 meter band

E1A08 (B) [97.219]

If a station in a message forwarding system inadvertently forwards a message that is in violation of FCC rules, who is primarily accountable for the rules violation?

B. The control operator of the originating station

E1A09 (A) [97.219]

What is the first action you should take if your digital message forwarding station inadvertently forwards a communication that violates FCC rules?

A. Discontinue forwarding the communication as soon as you become aware of it

E1A10 (A) [97.11]

If an amateur station is installed aboard a ship or aircraft, what condition must be met before the station is operated?

A. Its operation must be approved by the master of the ship or the pilot in command of the aircraft

E1A11 (B) [97.5]

What authorization or licensing is required when operating an amateur station aboard a US-registered vessel in international waters?

B. Any FCC-issued amateur license or a reciprocal permit for an alien amateur licensee

E1A12 (C) [97.301, 97.305]

With your transceiver displaying the carrier frequency of CW signals, you hear a DX station's CQ on 3.500 MHz. Is it legal to return the call using CW on the same frequency?

C. No, sidebands from the CW signal will be out of the band.

E1A13 (B) [97.5]

Who must be in physical control of the station apparatus of an amateur station aboard any vessel or craft that is documented or registered in the United States?

B. Any person holding an FCC-issued amateur license or who is authorized for alien reciprocal operation

E1B Station restrictions and special operations: restrictions on station location; general operating restrictions; spurious emissions, control operator reimbursement; antenna structure restrictions; RACES operations

E1B01 (D) [97.3]

Which of the following constitutes a spurious emission?

D. An emission outside its necessary bandwidth that can be reduced or eliminated without affecting the information transmitted

E1B02 (D) [97.13]

Which of the following factors might cause the physical location of an amateur station apparatus or antenna structure to be restricted?

D. The location is of environmental importance or significant in American history, architecture, or culture

E1B03 (A) [97.13]

Within what distance must an amateur station protect an FCC monitoring facility from harmful interference?

A. 1 mile

E1B04 (C) [97.13, 1.1305-1.1319]

What must be done before placing an amateur station within an officially designated wilderness area or wildlife preserve, or an area listed in the National Register of Historical Places?

C. An Environmental Assessment must be submitted to the FCC

E1B05 (D) [97.303]

What is the maximum bandwidth for a data emission on 60 meters?

D. 2.8 kHz

E1B06 (A) [97.15]

Which of the following additional rules apply if you are installing an amateur station antenna at a site at or near a public use airport?

A. You may have to notify the Federal Aviation Administration and register it with the FCC as required by Part 17 of FCC rules

E1B07 (B) [97.15]

Where must the carrier frequency of a CW signal be set to comply with FCC rules for 60 meter operation?

B. At the center frequency of the channel

E1B08 (D) [97.121]

What limitations may the FCC place on an amateur station if its signal causes interference to domestic broadcast reception, assuming that the receiver(s) involved are of good engineering design?

D. The amateur station must avoid transmitting during certain hours on frequencies that cause the interference

E1B09 (C) [97.407]

Which amateur stations may be operated in RACES?

C. Any FCC-licensed amateur station certified by the responsible civil defense organization for the area served

E1B10 (A) [97.407]

What frequencies are authorized to an amateur station participating in RACES?

A. All amateur service frequencies authorized to the control operator

E1B11 (A) [97.307]

What is the permitted mean power of any spurious emission relative to the mean power of the fundamental emission from a station transmitter or external RF amplifier installed after January 1, 2003, and transmitting on a frequency below 30 MHZ?

A. At least 43 dB below

E1B12 (B) [97.307]

What is the highest modulation index permitted at the highest modulation frequency for angle modulation?

B. 1.0

E1C Station Control: Definitions and restrictions pertaining to local, automatic and remote control operation; control operator responsibilities for remote and automatically controlled stations

E1C01 (D) [97.3]

What is a remotely controlled station?

D. A station controlled indirectly through a control link

E1C02 (A) [97.3, 97.109]

What is meant by automatic control of a station?

A. The use of devices and procedures for control so that the control operator does not have to be present at a control point

E1C03 (B) [97.3, 97.109]

How do the control operator responsibilities of a station under automatic control differ from one under local control?

B. Under automatic control the control operator is not required to be present at the control point

E1C04 (B) [97.109]

When may an automatically controlled station retransmit third party communications?

B. Only when transmitting RTTY or data emissions

E1C05 (A) [97.109]

When may an automatically controlled station originate third party communications?

A. Never

E1C06 (C) [97.109]

Which of the following statements concerning remotely controlled amateur stations is true?

C. A control operator must be present at the control point

E1C07 (C) [97.3]

What is meant by local control?

C. Direct manipulation of the transmitter by a control operator

E1C08 (B) [97.213]

What is the maximum permissible duration of a remotely controlled station's transmissions if its control link malfunctions?

B. 3 minutes

E1C09 (D) [97.205]

Which of these frequencies are available for an automatically controlled repeater operating below 30 MHz?

D. 29.500 - 29.700 MHz

E1C10 (B) [97.113]

What types of amateur stations may automatically retransmit the radio signals of other amateur stations?

B. Only auxiliary, repeater or space stations

E1D Amateur Satellite service: definitions and purpose; license requirements for space stations; available frequencies and bands; telecommand and telemetry operations; restrictions and special provisions; notification requirements

E1D01 (A) [97.3]

What is the definition of the term telemetry?

A. One-way transmission of measurements at a distance from the measuring instrument

E1D02 (C) [97.3]

What is the amateur satellite service?

C. A radio communications service using amateur radio stations on satellites

E1D03 (B) [97.3]

What is a telecommand station in the amateur satellite service?

B. An amateur station that transmits communications to initiate, modify or terminate functions of a space station

E1D04 (A) [97.3]

What is an Earth station in the amateur satellite service?

A. An amateur station within 50 km of the Earth's surface intended for communications with amateur stations by means of objects in space

E1D05 (C) [97.207]

What class of licensee is authorized to be the control operator of a space station?

C. All classes

E1D06 (A) [97.207]

Which of the following special provisions must a space station incorporate in order to comply with space station requirements?

A. The space station must be capable of terminating transmissions by telecommand when directed by the FCC

E1D07 (A) [97.207]

Which amateur service HF bands have frequencies authorized to space stations?

A. Only 40m, 20m, 17m, 15m, 12m and 10m

E1D08 (D) [97.207]

Which VHF amateur service bands have frequencies available for space stations?

D.   2 meters

E1D09 (B) [97.207]

Which amateur service UHF bands have frequencies available for a space station?

B. 70 cm, 23 cm, 13 cm

E1D10 (B) [97.211]

Which amateur stations are eligible to be telecommand stations?

B. Any amateur station so designated by the space station licensee, subject to the privileges of the class of operator license held by the control operator

E1D11 (D) [97.209]

Which amateur stations are eligible to operate as Earth stations?

D. Any amateur station, subject to the privileges of the class of operator license held by the control operator

E1E Volunteer examiner program: definitions; qualifications; preparation and administration of exams; accreditation; question pools; documentation requirements

E1E01 (D) [97.509]

What is the minimum number of qualified VEs required to administer an Element 4 amateur operator license examination?

D. 3

E1E02 (C) [97.523]

Where are the questions for all written US amateur license examinations listed?

C. In a question pool maintained by all the VECs

E1E03 (C) [97.521]

What is a Volunteer Examiner Coordinator?

C. An organization that has entered into an agreement with the FCC to coordinate amateur operator license examinations

E1E04 (D) [97.509, 97.525]

Which of the following best describes the Volunteer Examiner accreditation process?

D. The procedure by which a VEC confirms that the VE applicant meets FCC requirements to serve as an examiner

E1E05 (B) [97.503]

What is the minimum passing score on amateur operator license examinations?

B. Minimum passing score of 74%

E1E06 (C) [97.509]

Who is responsible for the proper conduct and necessary supervision during an amateur operator license examination session?

C. Each administering VE

E1E07 (B) [97.509]

What should a VE do if a candidate fails to comply with the examiner's instructions during an amateur operator license examination?

B. Immediately terminate the candidate's examination

E1E08 (C) [97.509]

To which of the following examinees may a VE not administer an examination?

C. Relatives of the VE as listed in the FCC rules

E1E09 (A) [97.509]

What may be the penalty for a VE who fraudulently administers or certifies an examination?

A. Revocation of the VE's amateur station license grant and the suspension of the VE's amateur operator license grant

E1E10 (C) [97.509]

What must the administering VEs do after the administration of a successful examination for an amateur operator license?

C. They must submit the application document to the coordinating VEC according to the coordinating VEC instructions

E1E11 (B) [97.509]

What must the VE team do if an examinee scores a passing grade on all examination elements needed for an upgrade or new license?

B. Three VEs must certify that the examinee is qualified for the license grant and that they have complied with the administering VE requirements

E1E12 (A) [97.509]

What must the VE team do with the application form if the examinee does not pass the exam?

A. Return the application document to the examinee

E1E13 (A) [97.519]

What are the consequences of failing to appear for re-administration of an examination when so directed by the FCC?

A. The licensee's license will be cancelled

E1E14 (A) [97.527]

For which types of out-of-pocket expenses do the Part 97 rules state that VEs and VECs may be reimbursed?

A. Preparing, processing, administering and coordinating an examination for an amateur radio license

E1F Miscellaneous rules: external RF power amplifiers; national quiet zone; business communications; compensated communications; spread spectrum; auxiliary stations; reciprocal operating privileges; IARP and CEPT licenses; third party communications with foreign countries; special temporary authority

E1F01 (B) [97.305]

On what frequencies are spread spectrum transmissions permitted?

B. Only on amateur frequencies above 222 MHz

E1F02 (A) [97.5]

Which of the following operating arrangements allows an FCC-licensed US citizen to operate in many European countries, and alien amateurs from many European countries to operate in the US?

A. CEPT agreement

E1F03 (A) [97.315]

Under what circumstances may a dealer sell an external RF power amplifier capable of operation below 144 MHz if it has not been granted FCC certification?

A. It was purchased in used condition from an amateur operator and is sold to another amateur operator for use at that operator's station

E1F04 (A) [97.3]

Which of the following geographic descriptions approximately describes "Line A"?

A. A line roughly parallel to and south of the US-Canadian border

E1F05 (D) [97.303]

Amateur stations may not transmit in which of the following frequency segments if they are located in the contiguous 48 states and north of Line A?

D. 420 - 430 MHz

E1F06 (C) [97.3]

What is the National Radio Quiet Zone?

C. An area surrounding the National Radio Astronomy Observatory

E1F07 (D) [97.113]

When may an amateur station send a message to a business?

D. When neither the amateur nor his or her employer has a pecuniary interest in the communications

E1F08 (A) [97.113]

Which of the following types of amateur station communications are prohibited?

A. Communications transmitted for hire or material compensation, except as otherwise provided in the rules

E1F09 (D) [97.311]

Which of the following conditions apply when transmitting spread spectrum emission?

D. All of these choices are correct

E1F10 (C) [97.313]

What is the maximum transmitter power for an amateur station transmitting spread spectrum communications?

C. 10 W

E1F11 (D) [97.317]

Which of the following best describes one of the standards that must be met by an external RF power amplifier if it is to qualify for a grant of FCC certification?

D. It must satisfy the FCC's spurious emission standards when operated at the lesser of 1500 watts, or its full output power

E1F12 (B) [97.201]

Who may be the control operator of an auxiliary station?

B. Only Technician, General, Advanced or Amateur Extra Class operators

E1F13 (C) [97.117]

What types of communications may be transmitted to amateur stations in foreign countries?

C. Communications incidental to the purpose of the amateur service and remarks of a personal nature

E1F14 (A) [1.931]

Under what circumstances might the FCC issue a "Special Temporary Authority" (STA) to an amateur station?

A. To provide for experimental amateur communications

SUBELEMENT E2 - OPERATING PROCEDURES [5 Exam Questions - 5 Groups]

E2A Amateur radio in space: amateur satellites; orbital mechanics; frequencies and modes; satellite hardware; satellite operations

E2A01 (C)

What is the direction of an ascending pass for an amateur satellite?

C. From south to north

E2A02 (A)

What is the direction of a descending pass for an amateur satellite?

A. From north to south

E2A03 (C)

What is the orbital period of an Earth satellite?

C. The time it takes for a satellite to complete one revolution around the Earth

E2A04 (B)

What is meant by the term mode as applied to an amateur radio satellite?

B. The satellite's uplink and downlink frequency bands

E2A05 (D)

What do the letters in a satellite's mode designator specify?

D. The uplink and downlink frequency ranges

E2A06 (A)

On what band would a satellite receive signals if it were operating in mode U/V?

A. 435-438 MHz

E2A07 (D)

Which of the following types of signals can be relayed through a linear transponder?

A. FM – CW

B. SSB – SSTV

C. PSK - Packet

D. All of these choices are correct

E2A08 (B)

Why should effective radiated power to a satellite which uses a linear transponder be limited?

B. To avoid reducing the downlink power to all other users

E2A09 (A)

What do the terms L band and S band specify with regard to satellite communications?

A. The 23 centimeter and 13 centimeter bands

E2A10 (A)

Why may the received signal from an amateur satellite exhibit a rapidly repeating fading effect?

A. Because the satellite is spinning

E2A11 (B)

What type of antenna can be used to minimize the effects of spin modulation and Faraday rotation?

B. A circularly polarized antenna

E2A12 (D)

What is one way to predict the location of a satellite at a given time?

D. By calculations using the Keplerian elements for the specified satellite

E2A13 (B)

What type of satellite appears to stay in one position in the sky?

B. Geostationary

E2B Television practices: fast scan television standards and techniques; slow scan television standards and techniques

E2B01 (A)

How many times per second is a new frame transmitted in a fast-scan (NTSC) television system?

A. 30

E2B02 (C)

How many horizontal lines make up a fast-scan (NTSC) television frame?

C. 525

E2B03 (D)

How is an interlaced scanning pattern generated in a fast-scan (NTSC) television system?

D. By scanning odd numbered lines in one field and even numbered ones in the next

E2B04 (B)

What is blanking in a video signal?

B. Turning off the scanning beam while it is traveling from right to left or from bottom to top

E2B05 (C)

Which of the following is an advantage of using vestigial sideband for standard fast-scan TV transmissions?

C. Vestigial sideband reduces bandwidth while allowing for simple video detector circuitry

E2B06 (A)

What is vestigial sideband modulation?

A. Amplitude modulation in which one complete sideband and a portion of the other are transmitted

E2B07 (B)

What is the name of the signal component that carries color information in NTSC video?

B. Chroma

E2B08 (D)

Which of the following is a common method of transmitting accompanying audio with amateur fast-scan television?

A. Frequency modulated sub=carrier

B. A separate VHF or UHF audio link

C. Frequency modulation of a video sub-carrier

D. All of these choices are correct

E2B09 (D)

What hardware, other than a receiver with SSB capability and a suitable computer, is needed to decode SSTV using Digital Radio Mondiale (DRM)?

D. No other hardware is needed

E2B10 (A)

Which of the following is an acceptable bandwidth for Digital Radio Mondiale (DRM) based voice or SSTV digital transmissions made on the HF amateur bands?

A. 3 KHz

E2B11 (B)

What is the function of the Vertical Interval Signaling (VIS) code transmitted as part of an SSTV transmission?

B. To identify the SSTV mode being used

E2B12 (D)

How are analog SSTV images typically transmitted on the HF bands?

D. Varying tone frequencies representing the video are transmitted using single sideband

E2B13 (C)

How many lines are commonly used in each frame on an amateur slow-scan color television picture?

C. 128 or 256

E2B14 (A)

What aspect of an amateur slow-scan television signal encodes the brightness of the picture?

A. Tone frequency

E2B15 (A)

What signals SSTV receiving equipment to begin a new picture line?

A. Specific tone frequencies

E2B16 (D)

Which of the following is the video standard used by North American Fast Scan ATV stations?

D. NTSC

E2B17 (B)

What is the approximate bandwidth of a slow-scan TV signal?

B. 3 kHz

E2B18 (D)

On which of the following frequencies is one likely to find FM ATV transmissions?

D. 1255 MHz

E2B19 (C)

What special operating frequency restrictions are imposed on slow scan TV transmissions?

C. They are restricted to phone band segments and their bandwidth can be no greater than that of a voice signal of the same modulation type

E2C Operating methods: contest and DX operating; spread-spectrum transmissions; selecting an operating frequency

E2C01 (A)

Which of the following is true about contest operating?

A. Operators are permitted to make contacts even if they do not submit a log

E2C02 (A)

Which of the following best describes the term "self-spotting" in regards to contest operating?

A. The generally prohibited practice of posting one's own call sign and frequency on a call sign spotting network

E2C03 (A)

From which of the following bands is amateur radio contesting generally excluded?

A. 30 meters

E2C04 (D)

On which of the following frequencies is an amateur radio contest contact generally discouraged?

D. 146.52 MHz

E2C05 (B)

What is the function of a DX QSL Manager?

B. To handle the receiving and sending of confirmation cards for a DX station

E2C06 (C)

During a VHF/UHF contest, in which band segment would you expect to find the highest level of activity?

C. In the weak signal segment of the band, with most of the activity near the calling frequency

E2C07 (A)

What is the Cabrillo format?

A. A standard for submission of electronic contest logs

E2C08 (A)

Why are received spread-spectrum signals resistant to interference?

A. Signals not using the spectrum-spreading algorithm are suppressed in the receiver

E2C09 (D)

How does the spread-spectrum technique of frequency hopping work?

D. The frequency of the transmitted signal is changed very rapidly according to a particular sequence also used by the receiving station

E2C10 (D)

Why might a DX station state that they are listening on another frequency?

A. Because the DX station may be transmitting on a frequency that is prohibited to some responding stations

B. To separate the calling stations from the DX station

C. To reduce interference, thereby improving operating efficiency

D. All of these choices are correct

E2C11 (A)

How should you generally identify your station when attempting to contact a DX station working a pileup or in a contest?

A. Send your full call sign once or twice

E2C12 (B)

What might help to restore contact when DX signals become too weak to copy across an entire HF band a few hours after sunset?

B. Switch to a lower frequency HF band

E2D Operating methods: VHF and UHF digital modes; APRS

E2D01 (B)

Which of the following digital modes is especially designed for use for meteor scatter signals?

B. FSK441

E2D02 (A)

What is the definition of baud?

A. The number of data symbols transmitted per second

E2D03 (D)

Which of the following digital modes is especially useful for EME communications?

D. JT65

E2D04 (C)

What is the purpose of digital store-and-forward functions on an Amateur Radio satellite?

E.   To store digital messages in the satellite for later download by other stations

E2D05 (B)

Which of the following techniques is normally used by low Earth orbiting digital satellites to relay messages around the world?

B. Store-and-forward

E2D06 (A)

Which of the following is a commonly used 2-meter APRS frequency?

A.   144.39 MHz

E2D07 (C)

Which of the following digital protocols is used by APRS?

C. AX.25

E2D08 (A)

Which of the following types of packet frames is used to transmit APRS beacon data?

A. Unnumbered Information

E2D09 (D)

Under clear communications conditions, which of these digital communications modes has the fastest data throughput?

D. 300-baud packet

E2D10 (C)

How can an APRS station be used to help support a public service communications activity?

C. An APRS station with a GPS unit can automatically transmit information to show a mobile station's position during the event

E2D11 (D)

Which of the following data are used by the APRS network communicate your location?

D. Latitude and longitude

E2D12 (A)

How does JT65 improve EME communications?

A. It can decode signals many dB below the noise floor using FEC

E2E Operating methods: operating HF digital modes; error correction

E2E01 (B)

Which type of modulation is common for data emissions below 30 MHz?

B. FSK

E2E02 (A)

What do the letters FEC mean as they relate to digital operation?

A. Forward Error Correction

E2E03 (C)

How is Forward Error Correction implemented?

C. By transmitting extra data that may be used to detect and correct transmission errors

E2E04 (A)

What is indicated when one of the ellipses in an FSK crossed-ellipse display suddenly disappears?

A. Selective fading has occurred

E2E05 (D)

How does ARQ accomplish error correction?

D. If errors are detected, a retransmission is requested

E2E06 (C)

What is the most common data rate used for HF packet communications?

C. 300 baud

E2E07 (B)

What is the typical bandwidth of a properly modulated MFSK16 signal?

B. 316 Hz

E2E08 (B)

Which of the following HF digital modes can be used to transfer binary files?

B. PACTOR

E2E09 (D)

Which of the following HF digital modes uses variable-length coding for bandwidth efficiency?

D. PSK31

E2E10 (C)

Which of these digital communications modes has the narrowest bandwidth?

C. PSK31

E2E11 (A)

What is the difference between direct FSK and audio FSK?

A. Direct FSK applies the data signal to the transmitter VFO

E2E12 (A)

Which type of digital communication does not support keyboard-to-keyboard operation?

A. Winlink

SUBELEMENT E3 - RADIO WAVE PROPAGATION [3 Exam Questions - 3 Groups]

E3A Propagation and technique: Earth-Moon-Earth communications (EME), meteor scatter

E3A01 (D)

What is the approximate maximum separation measured along the surface of the Earth between two stations communicating by Moon bounce?

D. 12,000 miles, as long as both can "see" the Moon

E3A02 (B)

What characterizes libration fading of an Earth-Moon-Earth signal?

B. A fluttery irregular fading

E3A03 (A)

When scheduling EME contacts, which of these conditions will generally result in the least path loss?

A. When the Moon is at perigee

E3A04 (D)

What type of receiving system is desirable for EME communications?

D. Equipment with very low noise figures

E3A05 (A)

Which of the following describes a method of establishing EME contacts?

A. Time synchronous transmissions with each station alternating

E3A06 (B)

What frequency range would you normally tune to find EME signals in the 2 meter band?

B. 144.000 - 144.100 MHz

E3A07 (D)

What frequency range would you normally tune to find EME signals in the 70 cm band?

D. 432.000 - 432.100 MHz

E3A08 (A)

When a meteor strikes the Earth's atmosphere, a cylindrical region of free electrons is formed at what layer of the ionosphere?

A. The E layer

E3A09 (C)

Which of the following frequency ranges is well suited for meteor-scatter communications?

C. 28 - 148 MHz

E3A10 (D)

Which of the following is a good technique for making meteor-scatter contacts?

  A. 15 second timed transmission sequences with stations alternating.
  B. Use of high speed CW or digital modes
  C. Short transmission with rapidly repeated call signs and signal reports

D. All of these choices are correct

E3B Propagation and technique: trans-equatorial, long path, gray-line; multi-path propagation

E3B01 (A)

What is transequatorial propagation?

A. Propagation between two mid-latitude points at approximately the same distance north and south of the magnetic equator

E3B02 (C)

What is the approximate maximum range for signals using transequatorial propagation?

C. 5000 miles

E3B03 (C)

What is the best time of day for transequatorial propagation?

C. Afternoon or early evening

E3B04 (A)

What type of propagation is probably occurring if an HF beam antenna must be pointed in a direction 180 degrees away from a station to receive the strongest signals?

A. Long-path

E3B05 (C)

Which amateur bands typically support long-path propagation?

C. 160 to 10 meters

E3B06 (B)

Which of the following amateur bands most frequently provides long-path propagation?

B. 20 meters

E3B07 (D)

Which of the following could account for hearing an echo on the received signal of a distant station?

D. Receipt of a signal by more than one path

E3B08 (D)

What type of HF propagation is probably occurring if radio signals travel along the terminator between daylight and darkness?

D. Gray-line

E3B09 (A)

At what time of day is gray-line propagation most likely to occur?

A. At sunrise and sunset

E3B10 (B)

What is the cause of gray-line propagation?

B. At twilight, D-layer absorption drops while E-layer and F-layer propagation remain strong

E3B11 (C)

Which of the following describes gray-line propagation?

C. Long distance communications at twilight on frequencies less than 15 MHz

E3C Propagation and technique: Aurora propagation selective fading; radio-path horizon; take-off angle over flat or sloping terrain; effects of ground on propagation; less common propagation modes

E3C01 (D)

Which of the following effects does Aurora activity have on radio communications?

A. SSB signals are raspy

B. Signals propagating through the Aurora are fluttery

C. CW signals appear to be modulated by white noise

D. All of these choices are correct

E3C02 (C)

What is the cause of Aurora activity?

C. The interaction of charged particles from the Sun with the Earth's magnetic field and the ionosphere

E3C03 (D)

Where in the ionosphere does Aurora activity occur?

D. In the E-region

E3C04 (A)

Which emission mode is best for Aurora propagation?

A. CW

E3C05 (B)

Which of the following describes selective fading?

B. Partial cancellation of some frequencies within the received pass band

E3C06 (A)

By how much does the VHF/UHF radio-path horizon distance exceed the geometric horizon?

A. By approximately 15% of the distance

E3C07 (B)

How does the radiation pattern of a horizontally polarized 3-element beam antenna vary with its height above ground?

B. The main lobe takeoff angle decreases with increasing height

E3C08 (B)

What is the name of the high-angle wave in HF propagation that travels for some distance within the F2 region?

B. Pedersen ray

E3C09 (C)

Which of the following is usually responsible for causing VHF signals to propagate for hundreds of miles?

C. Tropospheric ducting

E3C10 (B)

How does the performance of a horizontally polarized antenna mounted on the side of a hill compare with the same antenna mounted on flat ground?

B. The main lobe takeoff angle decreases in the downhill direction

E3C11 (B)

From the contiguous 48 states, in which approximate direction should an antenna be pointed to take maximum advantage of aurora propagation?

B. North

E3C12 (C)

How does the maximum distance of ground-wave propagation change when the signal frequency is increased?

C. It decreases

E3C13 (A)

What type of polarization is best for ground-wave propagation?

A. Vertical

E3C14 (D)

Why does the radio-path horizon distance exceed the geometric horizon?

F.   Downward bending due to density variations in the atmosphere

SUBELEMENT E4 - AMATEUR PRACTICES [5 Exam Questions - 5 Groups]

E4A Test equipment: analog and digital instruments; spectrum and network analyzers, antenna analyzers; oscilloscopes; testing transistors; RF measurements

E4A01 (C)

How does a spectrum analyzer differ from an oscilloscope?

C. A spectrum analyzer displays signals in the frequency domain; an oscilloscope displays signals in the time domain

E4A02 (D)

Which of the following parameters would a spectrum analyzer display on the horizontal axis?

D. Frequency

E4A03 (A)

Which of the following parameters would a spectrum analyzer display on the vertical axis?

A. Amplitude

E4A04 (A)

Which of the following test instruments is used to display spurious signals from a radio transmitter?

A. A spectrum analyzer

E4A05 (B)

Which of the following test instruments is used to display intermodulation distortion products in an SSB transmission?

B. A spectrum analyzer

E4A06 (D)

Which of the following could be determined with a spectrum analyzer?

A. The degree of isolation between the input and output ports of a 2 meter duplexer

B. Whether a crystal is operating on its fundamental or overtone frequency

C. The spectral output of a transmitter

D. All of these choices are correct

E4A07 (B)

Which of the following is an advantage of using an antenna analyzer compared to an SWR bridge to measure antenna SWR?

B. Antenna analyzers do not need an external RF source

E4A08 (D)

Which of the following instruments would be best for measuring the SWR of a beam antenna?

D. An antenna analyzer

E4A09 (A)

Which of the following describes a good method for measuring the intermodulation distortion of your own PSK signal?

A. Transmit into a dummy load, receive the signal on a second receiver, and feed the audio into the sound card of a computer running an appropriate PSK program

E4A10 (D)

Which of the following tests establishes that a silicon NPN junction transistor is biased on?

D. Measure base-to-emitter voltage with a voltmeter; it should be approximately 0.6 to 0.7 volts

E4A11 (B)

Which of these instruments could be used for detailed analysis of digital signals?

B. Oscilloscope

E4A12 (B)

Which of the following procedures is an important precaution to follow when connecting a spectrum analyzer to a transmitter output?

B. Attenuate the transmitter output going to the spectrum analyzer

E4B Measurement techniques: Instrument accuracy and performance limitations; probes; techniques to minimize errors; measurement of Q; instrument calibration

E4B01 (B)

Which of the following factors most affects the accuracy of a frequency counter?

B. Time base accuracy

E4B02 (C)

What is an advantage of using a bridge circuit to measure impedance?

C. The measurement is based on obtaining a signal null, which can be done very precisely

E4B03 (C)

If a frequency counter with a specified accuracy of +/- 1.0 ppm reads 146,520,000 Hz, what is the most the actual frequency being measured could differ from the reading?

C. 146.52 Hz

E4B04 (A)

If a frequency counter with a specified accuracy of +/- 0.1 ppm reads 146,520,000 Hz, what is the most the actual frequency being measured could differ from the reading?

A. 14.652 Hz

E4B05 (D)

If a frequency counter with a specified accuracy of +/- 10 ppm reads 146,520,000 Hz, what is the most the actual frequency being measured could differ from the reading?

D. 1465.20 Hz

E4B06 (D)

How much power is being absorbed by the load when a directional power meter connected between a transmitter and a terminating load reads 100 watts forward power and 25 watts reflected power?

D. 75 watts

E4B07 (A)

Which of the following is good practice when using an oscilloscope probe?

A. Keep the signal ground connection of the probe as short as possible

E4B08 (C)

Which of the following is a characteristic of a good DC voltmeter?

C. High impedance input

E4B09 (D)

What is indicated if the current reading on an RF ammeter placed in series with the antenna feed line of a transmitter increases as the transmitter is tuned to resonance?

E. There is more power going into the antenna

E4B10 (B)

Which of the following describes a method to measure intermodulation distortion in an SSB transmitter?

B. Modulate the transmitter with two non-harmonically related audio frequencies and observe the RF output with a spectrum analyzer

E4B11 (D)

How should a portable antenna analyzer be connected when measuring antenna resonance and feed point impedance?

D. Connect the antenna feed line directly to the analyzer's connector

E4B12 (A)

What is the significance of voltmeter sensitivity expressed in ohms per volt?

A. The full scale reading of the voltmeter multiplied by its ohms per volt rating will provide the input impedance of the voltmeter

E4B13 (A)

How is the compensation of an oscilloscope probe typically adjusted?

A. A square wave is displayed and the probe is adjusted until the horizontal portions of the displayed wave are as nearly flat as possible

E4B14 (B)

What happens if a dip meter is too tightly coupled to a tuned circuit being checked?

B. A less accurate reading results

E4B15 (C)

Which of the following can be used as a relative measurement of the Q for a series-tuned circuit?

C. The bandwidth of the circuit's frequency response

E4C Receiver performance characteristics: phase noise; capture effect; noise floor; image rejection; MDS; signal-to-noise-ratio; selectivity

E4C01 (D)

What is an effect of excessive phase noise in the local oscillator section of a receiver?

D. It can cause strong signals on nearby frequencies to interfere with reception of weak signals

E4C02 (A)

Which of the following portions of a receiver can be effective in eliminating image signal interference?

A. A front-end filter or pre-selector

E4C03 (C)

What is the term for the blocking of one FM phone signal by another, stronger FM phone signal?

C. Capture effect

E4C04 (D)

What is the definition of the noise figure of a receiver?

D. The ratio in dB of the noise generated by the receiver compared to the theoretical minimum noise

E4C05 (B)

What does a value of -174 dBm/Hz represent with regard to the noise floor of a receiver?

B. The theoretical noise at the input of a perfect receiver at room temperature

E4C06 (D)

A CW receiver with the AGC off has an equivalent input noise power density of -174 dBm/Hz. What would be the level of an unmodulated carrier input to this receiver that would yield an audio output SNR of 0 dB in a 400 Hz noise bandwidth?

D. -148 dBm

E4C07 (B)

What does the MDS of a receiver represent?

B. The minimum discernible signal

E4C08 (B)

How might lowering the noise figure affect receiver performance?

B. It would improve weak signal sensitivity

E4C09 (C)

Which of the following choices is a good reason for selecting a high frequency for the design of the IF in a conventional HF or VHF communications receiver?

C. Easier for front-end circuitry to eliminate image responses

E4C10 (B)

Which of the following is a desirable amount of selectivity for an amateur RTTY HF receiver?

B.   300 Hz

E4C11 (B)

Which of the following is a desirable amount of selectivity for an amateur SSB phone receiver?

B. 2.4 kHz

E4C12 (D)

What is an undesirable effect of using too wide a filter bandwidth in the IF section of a receiver?

D. Undesired signals may be heard

E4C13 (C)

How does a narrow-band roofing filter affect receiver performance?

C. It improves dynamic range by attenuating strong signals near the receive frequency

E4C14 (D)

On which of the following frequencies might a signal be transmitting which is generating a spurious image signal in a receiver tuned to 14.300 MHz and which uses a 455 kHz IF frequency?

D. 15.210 MHz

E4C15 (D)

What is the primary source of noise that can be heard from an HF receiver with an antenna connected?

D. Atmospheric noise

E4D Receiver performance characteristics: blocking dynamic range; intermodulation and cross-modulation interference; 3rd order intercept; desensitization; preselection

E4D01 (A)

What is meant by the blocking dynamic range of a receiver?

A. The difference in dB between the noise floor and the level of an incoming signal which will cause 1 dB of gain compression

E4D02 (A)

Which of the following describes two problems caused by poor dynamic range in a communications receiver?

A. Cross-modulation of the desired signal and desensitization from strong adjacent signals

E4D03 (B)

How can intermodulation interference between two repeaters occur?

B. When the repeaters are in close proximity and the signals mix in the final amplifier of one or both transmitters

E4D04 (B)

Which of the following may reduce or eliminate intermodulation interference in a repeater caused by another transmitter operating in close proximity?

B. A properly terminated circulator at the output of the transmitter

E4D05 (A)

What transmitter frequencies would cause an intermodulation-product signal in a receiver tuned to 146.70 MHz when a nearby station transmits on 146.52 MHz?

A. 146.34 MHz and 146.61 MHz

E4D06 (D)

What is the term for unwanted signals generated by the mixing of two or more signals?

D. Intermodulation interference

E4D07 (D)

Which of the following describes the most significant effect of an off-frequency signal when it is causing cross-modulation interference to a desired signal?

D. The off-frequency unwanted signal is heard in addition to the desired signal

E4D08 (C)

What causes intermodulation in an electronic circuit?

C. Nonlinear circuits or devices

E4D09 (C)

What is the purpose of the preselector in a communications receiver?

C. To increase rejection of unwanted signals

E4D10 (C)

What does a third-order intercept level of 40 dBm mean with respect to receiver performance?

C. A pair of 40 dBm signals will theoretically generate a third-order intermodulation product with the same level as the input signals

E4D11 (A)

Why are third-order intermodulation products created within a receiver of particular interest compared to other products?

A. The third-order product of two signals which are in the band of interest is also likely to be within the band

E4D12 (A)

What is the term for the reduction in receiver sensitivity caused by a strong signal near the received frequency?

A. Desensitization

E4D13 (B)

Which of the following can cause receiver desensitization?

B. Strong adjacent-channel signals

E4D14 (A)

Which of the following is a way to reduce the likelihood of receiver desensitization?

A. Decrease the RF bandwidth of the receiver

E4E Noise suppression: system noise; electrical appliance noise; line noise; locating noise sources; DSP noise reduction; noise blankers

E4E01 (A)

Which of the following types of receiver noise can often be reduced by use of a receiver noise blanker?

A. Ignition noise

E4E02 (D)

Which of the following types of receiver noise can often be reduced with a DSP noise filter?

A. Ignition noise

B. Broadband white noise

C. Heterodyne interference

D. All of these choices are correct

E4E03 (B)

Which of the following signals might a receiver noise blanker be able to remove from desired signals?

B. Signals which appear across a wide bandwidth

E4E04 (D)

How can conducted and radiated noise caused by an automobile alternator be suppressed?

D. By connecting the radio's power leads directly to the battery and by installing coaxial capacitors in line with the alternator leads

E4E05 (B)

How can noise from an electric motor be suppressed?

B. By installing a brute-force AC-line filter in series with the motor leads

E4E06 (B)

What is a major cause of atmospheric static?

B. Thunderstorms

E4E07 (C)

How can you determine if line noise interference is being generated within your home?

C. By turning off the AC power line main circuit breaker and listening on a battery operated radio

E4E08 (A)

What type of signal is picked up by electrical wiring near a radio antenna?

A. A common-mode signal at the frequency of the radio transmitter

E4E09 (C)

What undesirable effect can occur when using an IF noise blanker?

C. Nearby signals may appear to be excessively wide even if they meet emission standards

E4E10 (D)

What is a common characteristic of interference caused by a touch controlled electrical device?

A. The interfering signal sounds like AC hum on an AM receiver or a carrier modulated by 60 Hz hum on a SSB or CW receiver

B. The interfering signal may drift slowly across the HF spectrum

C. The interfering signal can be several kHz in width and usually repeats at regular intervals across a HF band

D. All of these choices are correct

E4E11 (B)

Which of the following is the most likely cause if you are hearing combinations of local AM broadcast signals within one or more of the MF or HF ham bands?

B. Nearby corroded metal joints are mixing and re-radiating the broadcast signals

E4E12 (A)

What is one disadvantage of using some types of automatic DSP notch-filters when attempting to copy CW signals?

A. The DSP filter can remove the desired signal at the same time as it removes interfering signals

E4E13 (D)

What might be the cause of a loud roaring or buzzing AC line interference that comes and goes at intervals?

A. Arcing contacts in a thermostatically controlled device

B. A defective doorbell or doorbell transformer inside a nearby residence

C. A malfunctioning illuminated advertising display

D. All of these choices are correct

E4E14 (C)

What is one type of electrical interference that might be caused by the operation of a nearby personal computer?

SUBELEMENT E5 - ELECTRICAL PRINCIPLES [4 Exam Questions - 4 Groups]

E5A Resonance and Q: characteristics of resonant circuits; series and parallel resonance; Q; half-power bandwidth; phase relationships in reactive circuits

E5A01 (A)

What can cause the voltage across reactances in series to be larger than the voltage applied to them?

A. Resonance

E5A02 (C)

What is resonance in an electrical circuit?

C. The frequency at which the capacitive reactance equals the inductive reactance

E5A03 (D)

What is the magnitude of the impedance of a series RLC circuit at resonance?

D. Approximately equal to circuit resistance

E5A04 (A)

What is the magnitude of the impedance of a circuit with a resistor, an inductor and a capacitor all in parallel, at resonance?

A. Approximately equal to circuit resistance

E5A05 (B)

What is the magnitude of the current at the input of a series RLC circuit as the frequency goes through resonance?

B. Maximum

E5A06 (B)

What is the magnitude of the circulating current within the components of a parallel LC circuit at resonance?

B. It is at a maximum

E5A07 (A)

What is the magnitude of the current at the input of a parallel RLC circuit at resonance?

A. Minimum

E5A08 (C)

What is the phase relationship between the current through and the voltage across a series resonant circuit at resonance?

C. The voltage and current are in phase

E5A09 (C)

What is the phase relationship between the current through and the voltage across a parallel resonant circuit at resonance?

C. The voltage and current are in phase

E5A10 (A)

What is the half-power bandwidth of a parallel resonant circuit that has a resonant frequency of 1.8 MHz and a Q of 95?

A. 18.9 kHz

E5A11 (C)

What is the half-power bandwidth of a parallel resonant circuit that has a resonant frequency of 7.1 MHz and a Q of 150?

C. 47.3 kHz

E5A12 (C)

What is the half-power bandwidth of a parallel resonant circuit that has a resonant frequency of 3.7 MHz and a Q of 118?

C. 31.4 kHz

E5A13 (B)

What is the half-power bandwidth of a parallel resonant circuit that has a resonant frequency of 14.25 MHz and a Q of 187?

B. 76.2 kHz

E5A14 (C)

What is the resonant frequency of a series RLC circuit if R is 22 ohms, L is 50 microhenrys and C is 40 picofarads?

C. 3.56 MHz

E5A15 (B)

What is the resonant frequency of a series RLC circuit if R is 56 ohms, L is 40 microhenrys and C is 200 picofarads?

B. 1.78 MHz

E5A16 (D)

What is the resonant frequency of a parallel RLC circuit if R is 33 ohms, L is 50 microhenrys and C is 10 picofarads?

C. 7.12 MHz

E5A17 (A)

What is the resonant frequency of a parallel RLC circuit if R is 47 ohms, L is 25 microhenrys and C is 10 picofarads?

A. 10.1 MHz

E5B Time constants and phase relationships: RLC time constants; definition; time constants in RL and RC circuits; phase angle between voltage and current; phase angles of series and parallel circuits

E5B01 (B)

What is the term for the time required for the capacitor in an RC circuit to be charged to 63.2% of the applied voltage?

B. One time constant

E5B02 (D)

What is the term for the time it takes for a charged capacitor in an RC circuit to discharge to 36.8% of its initial voltage?

D. One time constant

E5B03 (D)

The capacitor in an RC circuit is discharged to what percentage of the starting voltage after two time constants?

D. 13.5%

E5B04 (D)

What is the time constant of a circuit having two 220-microfarad capacitors and two 1-megohm resistors, all in parallel?

D. 220 seconds

E5B05 (A)

How long does it take for an initial charge of 20 V DC to decrease to 7.36 V DC in a 0.01-microfarad capacitor when a 2-megohm resistor is connected across it?

A. 0.02 seconds

E5B06 (C)

How long does it take for an initial charge of 800 V DC to decrease to 294 V DC in a 450-microfarad capacitor when a 1-megohm resistor is connected across it?

C. 450 seconds

E5B07 (C)

What is the phase angle between the voltage across and the current through a series RLC circuit if XC is 500 ohms, R is 1 kilohm, and XL is 250 ohms?

C. 14.0 degrees with the voltage lagging the current

E5B08 (A)

What is the phase angle between the voltage across and the current through a series RLC circuit if XC is 100 ohms, R is 100 ohms, and XL is 75 ohms?

A. 14 degrees with the voltage lagging the current

E5B09 (D)

What is the relationship between the current through a capacitor and the voltage across a capacitor?

D. Current leads voltage by 90 degrees

E5B10 (A)

What is the relationship between the current through an inductor and the voltage across an inductor?

A. Voltage leads current by 90 degrees

E5B11 (B)

What is the phase angle between the voltage across and the current through a series RLC circuit if XC is 25 ohms, R is 100 ohms, and XL is 50 ohms?

B. 14 degrees with the voltage leading the current

E5B12 (C)

What is the phase angle between the voltage across and the current through a series RLC circuit if XC is 75 ohms, R is 100 ohms, and XL is 50 ohms?

C. 14 degrees with the voltage lagging the current

E5B13 (D)

What is the phase angle between the voltage across and the current through a series RLC circuit if XC is 250 ohms, R is 1 kilohm, and XL is 500 ohms?

D. 14.04 degrees with the voltage leading the current

E5C Impedance plots and coordinate systems: plotting impedances in polar coordinates; rectangular coordinates

E5C01 (B)

In polar coordinates, what is the impedance of a network consisting of a 100-ohm-reactance inductor in series with a 100-ohm resistor?

B. 141 ohms at an angle of 45 degrees

E5C02 (D)

In polar coordinates, what is the impedance of a network consisting of a 100-ohm-reactance inductor, a 100-ohm-reactance capacitor, and a 100-ohm resistor, all connected in series?

D. 100 ohms at an angle of 0 degrees

E5C03 (A)

In polar coordinates, what is the impedance of a network consisting of a 300-ohm-reactance capacitor, a 600-ohm-reactance inductor, and a 400-ohm resistor, all connected in series?

A. 500 ohms at an angle of 37 degrees

E5C04 (D)

In polar coordinates, what is the impedance of a network consisting of a 400-ohm-reactance capacitor in series with a 300-ohm resistor?

D. 500 ohms at an angle of -53.1 degrees

E5C05 (A)

In polar coordinates, what is the impedance of a network consisting of a 400-ohm-reactance inductor in parallel with a 300-ohm resistor?

A. 240 ohms at an angle of 36.9 degrees

E5C06 (D)

In polar coordinates, what is the impedance of a network consisting of a 100-ohm-reactance capacitor in series with a 100-ohm resistor?

D. 141 ohms at an angle of -45 degrees

E5C07 (C)

In polar coordinates, what is the impedance of a network comprised of a 100-ohm-reactance capacitor in parallel with a 100-ohm resistor?

C. 71 ohms at an angle of -45 degrees

E5C08 (B)

In polar coordinates, what is the impedance of a network comprised of a 300-ohm-reactance inductor in series with a 400-ohm resistor?

B. 500 ohms at an angle of 37 degrees

E5C09 (A)

When using rectangular coordinates to graph the impedance of a circuit, what does the horizontal axis represent?

A. Resistive component

E5C10 (B)

When using rectangular coordinates to graph the impedance of a circuit, what does the vertical axis represent?

E5C11 (C)

What do the two numbers represent that are used to define a point on a graph using rectangular coordinates?

C. The coordinate values along the horizontal and vertical axes

E5C12 (D)

If you plot the impedance of a circuit using the rectangular coordinate system and find the impedance point falls on the right side of the graph on the horizontal axis, what do you know about the circuit?

D. It is equivalent to a pure resistance

E5C13 (D)

What coordinate system is often used to display the resistive, inductive, and/or capacitive reactance components of an impedance?

D. Rectangular coordinates

E5C14 (D)

What coordinate system is often used to display the phase angle of a circuit containing resistance, inductive and/or capacitive reactance?

D. Polar coordinates

E5C15 (A)

In polar coordinates, what is the impedance of a circuit of 100 -j100 ohms impedance?

A. 141 ohms at an angle of -45 degrees

E5C16 (B)

In polar coordinates, what is the impedance of a circuit that has an admittance of 7.09 millisiemens at 45 degrees?

B. 141 ohms at an angle of -45 degrees

E5C17 (C)

In rectangular coordinates, what is the impedance of a circuit that has an admittance of 5 millisiemens at -30 degrees?

C. 173 +j100 ohms

E5C18 (B)

In polar coordinates, what is the impedance of a series circuit consisting of a resistance of 4 ohms, an inductive reactance of 4 ohms, and a capacitive reactance of 1 ohm?

C. 5 ohms at an angle of 37 degrees

E5C19 (B)

Which point on Figure E5-2 best represents that impedance of a series circuit consisting of a 400 ohm resistor and a 38 picofarad capacitor at 14 MHz?

B. Point 4

E5C20 (B)

Which point in Figure E5-2 best represents the impedance of a series circuit consisting of a 300 ohm resistor and an 18 microhenry inductor at 3.505 MHz?

B. Point 3

E5C21 (A)

Which point on Figure E5-2 best represents the impedance of a series circuit consisting of a 300 ohm resistor and a 19 picofarad capacitor at 21.200 MHz?

A. Point 1

E5C22 (A)

In rectangular coordinates, what is the impedance of a network consisting of a 10-microhenry inductor in series with a 40-ohm resistor at 500 MHz?

A. 40 + j31,400

E5C23 (D)

Which point on Figure E5-2 best represents the impedance of a series circuit consisting of a 300-ohm resistor, a 0.64-microhenry inductor and an 85-picofarad capacitor at 24.900 MHz?

D. Point 8

E5D AC and RF energy in real circuits: skin effect; electrostatic and electromagnetic fields; reactive power; power factor; coordinate systems

E5D01 (A)

What is the result of skin effect?

A. As frequency increases, RF current flows in a thinner layer of the conductor, closer to the surface

E5D02 (C)

Why is the resistance of a conductor different for RF currents than for direct currents?

C. Because of skin effect

E5D03 (C)

What device is used to store electrical energy in an electrostatic field?

C. A capacitor

E5D04 (B)

What unit measures electrical energy stored in an electrostatic field?

B. Joule

E5D05 (B)

Which of the following creates a magnetic field?

B. Electric current

E5D06 (D)

In what direction is the magnetic field oriented about a conductor in relation to the direction of electron flow?

D. In a direction determined by the left-hand rule

E5D07 (D)

What determines the strength of a magnetic field around a conductor?

D. The amount of current

E5D08 (B)

What type of energy is stored in an electromagnetic or electrostatic field?

B. Potential energy

E5D09 (B)

What happens to reactive power in an AC circuit that has both ideal inductors and ideal capacitors?

B. It is repeatedly exchanged between the associated magnetic and electric fields, but is not dissipated

E5D10 (A)

How can the true power be determined in an AC circuit where the voltage and current are out of phase?

A. By multiplying the apparent power times the power factor

E5D11 (C)

What is the power factor of an R-L circuit having a 60 degree phase angle between the voltage and the current?

C. 0.5

E5D12 (B)

How many watts are consumed in a circuit having a power factor of 0.2 if the input is 100-V AC at 4 amperes?

B. 80 watts

E5D13 (B)

How much power is consumed in a circuit consisting of a 100 ohm resistor in series with a 100 ohm inductive reactance drawing 1 ampere?

B. 100 Watts

E5D14 (A)

What is reactive power?

A. Wattless, nonproductive power

E5D15 (D)

What is the power factor of an RL circuit having a 45 degree phase angle between the voltage and the current?

D. 0.707

E5D16 (C)

What is the power factor of an RL circuit having a 30 degree phase angle between the voltage and the current?

C. 0.866

E5D17 (D)

How many watts are consumed in a circuit having a power factor of 0.6 if the input is 200V AC at 5 amperes?

D. 600 watts

E5D18 (B)

How many watts are consumed in a circuit having a power factor of 0.71 if the apparent power is 500 VA?

B. 355 W

SUBELEMENT E6 - CIRCUIT COMPONENTS [6 Exam Questions - 6 Groups]

E6A Semiconductor materials and devices: semiconductor materials; germanium, silicon, P-type, N-type; transistor types: NPN, PNP, junction, field-effect transistors: enhancement mode; depletion mode; MOS; CMOS; N-channel; P-channel

E6A01 (C)

In what application is gallium arsenide used as a semiconductor material in preference to germanium or silicon?

C. At microwave frequencies

E6A02 (A)

Which of the following semiconductor materials contains excess free electrons?

A. N-type

E6A03 (C)

What are the majority charge carriers in P-type semiconductor material?

C. Holes

E6A04 (C)

What is the name given to an impurity atom that adds holes to a semiconductor crystal structure?

C. Acceptor impurity

E6A05 (C)

What is the alpha of a bipolar junction transistor?

C. The change of collector current with respect to emitter current

E6A06 (B)

What is the beta of a bipolar junction transistor?

B. The change in collector current with respect to base current

E6A07 (A)

In Figure E6-1, what is the schematic symbol for a PNP transistor?

A. 1

E6A08 (D)

What term indicates the frequency at which the grounded-base current gain of a transistor has decreased to 0.7 of the gain obtainable at 1 kHz?

D. Alpha cutoff frequency

E6A09 (A)

What is a depletion-mode FET?

A. An FET that exhibits a current flow between source and drain when no gate voltage is applied

E6A10 (B)

In Figure E6-2, what is the schematic symbol for an N-channel dual-gate MOSFET?

B. 4

E6A11 (A)

In Figure E6-2, what is the schematic symbol for a P-channel junction FET?

A. 1

E6A12 (D)

Why do many MOSFET devices have internally connected Zener diodes on the gates?

D. To reduce the chance of the gate insulation being punctured by static discharges or excessive voltages

E6A13 (C)

What do the initials CMOS stand for?

C. Complementary Metal-Oxide Semiconductor

E6A14 (C)

How does DC input impedance at the gate of a field-effect transistor compare with the DC input impedance of a bipolar transistor?

C. An FET has high input impedance; a bipolar transistor has low input impedance

E6A15 (B)

Which of the following semiconductor materials contains an excess of holes in the outer shell of electrons?

B. P-type

E6A16 (B)

What are the majority charge carriers in N-type semiconductor material?

B. Free electrons

E6A17 (D)

What are the names of the three terminals of a field-effect transistor?

D. Gate, drain, source

E6B Semiconductor diodes

E6B01 (B)

What is the most useful characteristic of a Zener diode?

B. A constant voltage drop under conditions of varying current

E6B02 (D)

What is an important characteristic of a Schottky diode as compared to an ordinary silicon diode when used as a power supply rectifier?

D. Less forward voltage drop

E6B03 (C)

What special type of diode is capable of both amplification and oscillation?

C. Tunnel

E6B04 (A)

What type of semiconductor device is designed for use as a voltage-controlled capacitor?

A. Varactor diode

E6B05 (D)

What characteristic of a PIN diode makes it useful as an RF switch or attenuator?

D. A large region of intrinsic material

E6B06 (D)

Which of the following is a common use of a hot-carrier diode?

D. As a VHF / UHF mixer or detector

E6B07 (B)

What is the failure mechanism when a junction diode fails due to excessive current?

C. Excessive junction temperature

E6B08 (A)

Which of the following describes a type of semiconductor diode?

A. Metal-semiconductor junction

E6B09 (C)

What is a common use for point contact diodes?

C. As an RF detector

E6B10 (B)

In Figure E6-3, what is the schematic symbol for a light-emitting diode?

B. 5

E6B11 (A)

What is used to control the attenuation of RF signals by a PIN diode?

A. Forward DC bias current

E6B12 (C)

What is one common use for PIN diodes?

C. As an RF switch

E6B13 (B)

What type of bias is required for an LED to emit light?

B. Forward bias

E6C Integrated circuits: TTL digital integrated circuits; CMOS digital integrated circuits; gates

E6C01 (C)

What is the recommended power supply voltage for TTL series integrated circuits?

C. 5 volts

E6C02 (A)

What logic state do the inputs of a TTL device assume if they are left open?

A. A logic-high state

E6C03 (A)

Which of the following describes tri-state logic?

A. Logic devices with 0, 1, and high impedance output states

E6C04 (B)

Which of the following is the primary advantage of tri-state logic?

B. Ability to connect many device outputs to a common bus

E6C05 (D)

Which of the following is an advantage of CMOS logic devices over TTL devices?

D. Lower power consumption

E6C06 (C)

Why do CMOS digital integrated circuits have high immunity to noise on the input signal or power supply?

C. The input switching threshold is about one-half the power supply voltage

E6C07 (A)

In Figure E6-5, what is the schematic symbol for an AND gate?

A. 1

E6C08 (B)

In Figure E6-5, what is the schematic symbol for a NAND gate?

B. 2

E6C09 (B)

In Figure E6-5, what is the schematic symbol for an OR gate?

B. 3

E6C10 (D)

In Figure E6-5, what is the schematic symbol for a NOR gate?

D. 4

E6C11 (C)

In Figure E6-5, what is the schematic symbol for the NOT operation (inverter)?

C. 5

E6C12 (D)

What is BiCMOS logic?

D. An integrated circuit logic family using both bipolar and CMOS transistors

E6C13 (C)

Which of the following is an advantage of BiCMOS logic?

C. It has the high input impedance of CMOS and the low output impedance of bipolar transistors

E6D Optical devices and toroids: cathode-ray tube devices; charge-coupled devices (CCDs); liquid crystal displays (LCDs) Toroids: permeability; core material; selecting; winding

E6D01 (D)

What is cathode ray tube (CRT) persistence?

D. The length of time the image remains on the screen after the beam is turned off

E6D02 (B)

Exceeding what design rating can cause a cathode ray tube (CRT) to generate X-rays?

B. The anode voltage

E6D03 (C)

Which of the following is true of a charge-coupled device (CCD)?

C. It samples an analog signal and passes it in stages from the input to the output

E6D04 (A)

What function does a charge-coupled device (CCD) serve in a modern video camera?

A. It stores photogenerated charges as signals corresponding to pixels

E6D05 (B)

What is a liquid-crystal display (LCD)?

B. A display using a crystalline liquid which, in conjunction with polarizing filters, becomes opaque when voltage is applied

E6D06 (D)

What core material property determines the inductance of a toroidal inductor?

D. Permeability

E6D07 (B)

What is the usable frequency range of inductors that use toroidal cores, assuming a correct selection of core material for the frequency being used?

B. From less than 20 Hz to approximately 300 MHz

E6D08 (B)

What is one important reason for using powdered-iron toroids rather than ferrite toroids in an inductor?

B. Powdered-iron toroids generally maintain their characteristics at higher currents

E6D09 (C)

What devices are commonly used as VHF and UHF parasitic suppressors at the input and output terminals of transistorized HF amplifiers?

C. Ferrite beads

E6D10 (A)

What is a primary advantage of using a toroidal core instead of a solenoidal core in an inductor?

A. Toroidal cores confine most of the magnetic field within the core material

E6D11 (C)

How many turns will be required to produce a 1-mH inductor using a ferrite toroidal core that has an inductance index (A L) value of 523 millihenrys/1000 turns?

C. 43 turns

E6D12 (A)

How many turns will be required to produce a 5-microhenry inductor using a powdered-iron toroidal core that has an inductance index (A L) value of 40 microhenrys/100 turns?

A. 35 turns

E6D13 (D)

What type of CRT deflection is better when high-frequency waveforms are to be displayed on the screen?

D. Electrostatic

E6D14 (C)

Which is NOT true of a charge-coupled device (CCD)?

C. It is commonly used as an analog-to-digital converter

E6D15 (A)

What is the principle advantage of liquid-crystal display (LCD) devices over other types of display devices?

A. They consume less power

E6D16 (C)

What is one reason for using ferrite toroids rather than powdered-iron toroids in an inductor?

C. Ferrite toroids generally require fewer turns to produce a given inductance value

E6E Piezoelectric crystals and MMICs: quartz crystal oscillators and crystal filters); monolithic amplifiers

E6E01 (D)

What is a crystal lattice filter?

D. A filter with narrow bandwidth and steep skirts made using quartz crystals

E6E02 (A)

Which of the following factors has the greatest effect in helping determine the bandwidth and response shape of a crystal ladder filter?

A. The relative frequencies of the individual crystals

E6E03 (A)

What is one aspect of the piezoelectric effect?

A. Physical deformation of a crystal by the application of a voltage

E6E04 (A)

What is the most common input and output impedance of circuits that use MMICs?

A. 50 ohms

E6E05 (A)

Which of the following noise figure values is typical of a low-noise UHF preamplifier?

A. 2 dB

E6E06 (D)

What characteristics of the MMIC make it a popular choice for VHF through microwave circuits?

D. Controlled gain, low noise figure, and constant input and output impedance over the specified frequency range

E6E07 (B)

Which of the following is typically used to construct a MMIC-based microwave amplifier?

B. Microstrip construction

E6E08 (A)

How is power-supply voltage normally furnished to the most common type of monolithic microwave integrated circuit (MMIC)?

A. Through a resistor and/or RF choke connected to the amplifier output lead

E6E09 (B)

Which of the following must be done to insure that a crystal oscillator provides the frequency specified by the crystal manufacturer?

B. Provide the crystal with a specified parallel capacitance

E6E10 (A)

What is the equivalent circuit of a quartz crystal?

A. Motional capacitance, motional inductance and loss resistance in series, with a shunt capacitance representing electrode and stray capacitance

E6E11 (D)

Which of the following materials is likely to provide the highest frequency of operation when used in MMICs?

D. Gallium nitride

E6E12 (B)

What is a "Jones filter" as used as part of a HF receiver IF stage?

B. A variable bandwidth crystal lattice filter

E6F Optical components and power systems: photoconductive principles and effects, photovoltaic systems, optical couplers, optical sensors, and optoisolators

E6F01 (B)

What is photoconductivity?

B. The increased conductivity of an illuminated semiconductor

E6F02 (A)

What happens to the conductivity of a photoconductive material when light shines on it?

A. It increases

E6F03 (D)

What is the most common configuration of an optoisolator or optocoupler?

D. An LED and a phototransistor

E6F04 (B)

What is the photovoltaic effect?

B. The conversion of light to electrical energy

E6F05 (A)

Which of the following describes an optical shaft encoder?

A. A device which detects rotation of a control by interrupting a light source with a patterned wheel

E6F06 (A)

Which of these materials is affected the most by photoconductivity?

A. A crystalline semiconductor

E6F07 (B)

What is a solid state relay?

B. A device that uses semiconductor devices to implement the functions of an electromechanical relay

E6F08 (C)

Why are optoisolators often used in conjunction with solid state circuits when switching 120 VAC?

C. Optoisolators provide a very high degree of electrical isolation between a control circuit and the circuit being switched

E6F09 (D)

What is the efficiency of a photovoltaic cell?

D. The relative fraction of light that is converted to current

E6F10 (B)

What is the most common type of photovoltaic cell used for electrical power generation?

B. Silicon

E6F11 (B)

Which of the following is the approximate open-circuit voltage produced by a fully-illuminated silicon photovoltaic cell?

B. 0.5 V

E6F12 (C)

What absorbs the energy from light falling on a photovoltaic cell?

C. Electrons

SUBELEMENT E7 - PRACTICAL CIRCUITS [8 Exam Questions - 8 Groups]

E7A Digital circuits: digital circuit principles and logic circuits: classes of logic elements; positive and negative logic; frequency dividers; truth tables

E7A01 (C)

Which of the following is a bistable circuit?

C. A flip-flop

E7A02 (C)

How many output level changes are obtained for every two trigger pulses applied to the input of a T flip-flop circuit?

C. Two

E7A03 (B)

Which of the following can divide the frequency of a pulse train by 2?

B. A flip-flop

E7A04 (B)

How many flip-flops are required to divide a signal frequency by 4?

B. 2

E7A05 (D)

Which of the following is a circuit that continuously alternates between two states without an external clock?

D. Astable multivibrator

E7A06 (A)

What is a characteristic of a monostable multivibrator?

A. It switches momentarily to the opposite binary state and then returns, after a set time, to its original state

E7A07 (D)

What logical operation does a NAND gate perform?

D. It produces a logic "0" at its output only when all inputs are logic "1"

E7A08 (A)

What logical operation does an OR gate perform?

A. It produces a logic "1" at its output if any or all inputs are logic "1"

E7A09 (C)

What logical operation is performed by a two-input exclusive NOR gate?

C. It produces a logic "0" at its output if any single input is a logic "1"

E7A10 (C)

What is a truth table?

C. A list of inputs and corresponding outputs for a digital device

E7A11 (D)

What is the name for logic which represents a logic "1" as a high voltage?

D. Positive Logic

E7A12 (C)

What is the name for logic which represents a logic "0" as a high voltage?

C. Negative logic

E7A13 (B)

What is an SR or RS flip-flop?

B. A set/reset flip-flop whose output is low when R is high and S is low, high when S is high and R is low, and unchanged when both inputs are low

E7A14 (A)

What is a JK flip-flop?

A. A flip-flop similar to an RS except that it toggles when both J and K are high

E7A15 (A)

What is a D flip-flop?

A. A flip-flop whose output takes on the state of the D input when the clock signal transitions from low to high

E7B Amplifiers: Class of operation; vacuum tube and solid-state circuits; distortion and intermodulation; spurious and parasitic suppression; microwave amplifiers

E7B01 (A)

For what portion of a signal cycle does a Class AB amplifier operate?

A. More than 180 degrees but less than 360 degrees

E7B02 (A)

What is a Class D amplifier?

A. A type of amplifier that uses switching technology to achieve high efficiency

E7B03 (A)

Which of the following forms the output of a class D amplifier circuit?

A. A low-pass filter to remove switching signal components

E7B04 (A)

Where on the load line of a Class A common emitter amplifier would bias normally be set?

A. Approximately half-way between saturation and cutoff

E7B05 (C)

What can be done to prevent unwanted oscillations in an RF power amplifier?

C. Install parasitic suppressors and/or neutralize the stage

E7B06 (B)

Which of the following amplifier types reduces or eliminates even-order harmonics?

B. Push-pull

E7B07 (D)

Which of the following is a likely result when a Class C amplifier is used to amplify a single-sideband phone signal?

D. Signal distortion and excessive bandwidth

E7B08 (C)

How can an RF power amplifier be neutralized?

C. By feeding a 180-degree out-of-phase portion of the output back to the input

E7B09 (D)

Which of the following describes how the loading and tuning capacitors are to be adjusted when tuning a vacuum tube RF power amplifier that employs a pi-network output circuit?

D. The tuning capacitor is adjusted for minimum plate current, while the loading capacitor is adjusted for maximum permissible plate current

E7B10 (B)

In Figure E7-1, what is the purpose of R1 and R2?

B. Fixed bias

E7B11 (D)

In Figure E7-1, what is the purpose of R3?

D. Self bias

E7B12 (C)

What type of circuit is shown in Figure E7-1?

C. Common emitter amplifier

E7B13 (A)

In Figure E7-2, what is the purpose of R?

A. Emitter load

E7B14 (A)

In Figure E7-2, what is the purpose of C2?

A. Output coupling

E7B15 (C)

What is one way to prevent thermal runaway in a bipolar transistor amplifier?

C. Use a resistor in series with the emitter

E7B16 (A)

What is the effect of intermodulation products in a linear power amplifier?

A. Transmission of spurious signals

E7B17 (A)

Why are third-order intermodulation distortion products of particular concern in linear power amplifiers?

A. Because they are relatively close in frequency to the desired signal

E7B18 (C)

Which of the following is a characteristic of a grounded-grid amplifier?

C. Low input impedance

E7B19 (D)

What is a klystron?

D. A VHF, UHF, or microwave vacuum tube that uses velocity modulation

E7B20 (B)

What is a parametric amplifier?

B. A low-noise VHF or UHF amplifier relying on varying reactance for amplification

E7B21 (A)

Which of the following devices is generally best suited for UHF or microwave power amplifier applications?

A. Field effect transistor

E7C Filters and matching networks: types of networks; types of filters; filter applications; filter characteristics; impedance matching; DSP filtering

E7C01 (D)

How are the capacitors and inductors of a low-pass filter Pi-network arranged between the network's input and output?

D. A capacitor is connected between the input and ground, another capacitor is connected between the output and ground, and an inductor is connected between input and output

E7C02 (C)

A T-network with series capacitors and a parallel shunt inductor has which of the following properties?

C. It is a high-pass filter

E7C03 (A)

What advantage does a Pi-L-network have over a Pi-network for impedance matching between the final amplifier of a vacuum-tube transmitter and an antenna?

A. Greater harmonic suppression

E7C04 (C)

How does an impedance-matching circuit transform a complex impedance to a resistive impedance?

C. It cancels the reactive part of the impedance and changes the resistive part to a desired value

E7C05 (D)

Which filter type is described as having ripple in the passband and a sharp cutoff?

D. A Chebyshev filter

E7C06 (C)

What are the distinguishing features of an elliptical filter?

C. Extremely sharp cutoff with one or more notches in the stop band

E7C07 (B)

What kind of filter would you use to attenuate an interfering carrier signal while receiving an SSB transmission?

B. A notch filter

E7C08 (A)

What kind of digital signal processing audio filter might be used to remove unwanted noise from a received SSB signal?

A. An adaptive filter

E7C09 (C)

What type of digital signal processing filter might be used to generate an SSB signal?

C. A Hilbert-transform filter

E7C10 (B)

Which of the following filters would be the best choice for use in a 2 meter repeater duplexer?

B. A cavity filter

E7C11 (D)

Which of the following is the common name for a filter network which is equivalent to two L networks connected back-to-back with the inductors in series and the capacitors in shunt at the input and output?

D. Pi

E7C12 (B)

Which of the following describes a Pi-L network used for matching a vacuum-tube final amplifier to a 50-ohm unbalanced output?

B. A Pi network with an additional series inductor on the output

E7C13 (A)

What is one advantage of a Pi matching network over an L matching network consisting of a single inductor and a single capacitor?

A. The Q of Pi networks can be varied depending on the component values chosen

E7C14 (C)

Which of these modes is most affected by non-linear phase response in a receiver IF filter?

C. Digital

E7D Power supplies and voltage regulators

E7D01 (D)

What is one characteristic of a linear electronic voltage regulator?

D. The conduction of a control element is varied to maintain a constant output voltage

E7D02 (C)

What is one characteristic of a switching electronic voltage regulator?

C. The control device's duty cycle is controlled to produce a constant average output voltage

E7D03 (A)

What device is typically used as a stable reference voltage in a linear voltage regulator?

A. A Zener diode

E7D04 (B)

Which of the following types of linear voltage regulator usually make the most efficient use of the primary power source?

B. A series regulator

E7D05 (D)

Which of the following types of linear voltage regulator places a constant load on the unregulated voltage source?

D. A shunt regulator

E7D06 (C)

What is the purpose of Q1 in the circuit shown in Figure E7-3?

C. It increases the current-handling capability of the regulator

E7D07 (A)

What is the purpose of C2 in the circuit shown in Figure E7-3?

A. It bypasses hum around D1

E7D08 (C)

What type of circuit is shown in Figure E7-3?

C. Linear voltage regulator

E7D09 (D)

What is the purpose of C1 in the circuit shown in Figure E7-3?

D. It filters the supply voltage

E7D10 (A)

What is the purpose of C3 in the circuit shown in Figure E7-3?

A. It prevents self-oscillation

E7D11 (C)

What is the purpose of R1 in the circuit shown in Figure E7-3?

C. It supplies current to D1

E7D12 (D)

What is the purpose of R2 in the circuit shown in Figure E7-3?

D. It provides a constant minimum load for Q1

E7D13 (B)

What is the purpose of D1 in the circuit shown in Figure E7-3?

B. To provide a voltage reference

E7D14 (C)

What is one purpose of a "bleeder" resistor in a conventional (unregulated) power supply?

C. To improve output voltage regulation

E7D15 (D)

What is the purpose of a "step-start" circuit in a high-voltage power supply?

D. To allow the filter capacitors to charge gradually

E7D16 (D)

When several electrolytic filter capacitors are connected in series to increase the operating voltage of a power supply filter circuit, why should resistors be connected across each capacitor?

  A. To equalize, as much as possible, the voltage drop across each capacitor

  B. To provide a safety bleeder to discharge the capacitors when the supply is off

  C. To provide a minimum load current to reduce voltage excursions at light loads

  D. All of these choices are correct

E7D17 (C)

What is the primary reason that a high-frequency inverter type high-voltage power supply can be both less expensive and lighter in weight than a conventional power supply?

  C. The high frequency inverter design uses much smaller transformers and filter components for an equivalent power output

E7E Modulation and demodulation: reactance, phase and balanced modulators; detectors; mixer stages; DSP modulation and demodulation; software defined radio systems

E7E01 (B)

Which of the following can be used to generate FM phone emissions?

B. A reactance modulator on the oscillator

E7E02 (D)

What is the function of a reactance modulator?

D. To produce PM signals by using an electrically variable inductance or capacitance

E7E03 (C)

How does an analog phase modulator function?

C. By varying the tuning of an amplifier tank circuit to produce PM signals

E7E04 (A)

What is one way a single-sideband phone signal can be generated?

A. By using a balanced modulator followed by a filter

E7E05 (D)

What circuit is added to an FM transmitter to boost the higher audio frequencies?

D. A pre-emphasis network

E7E06 (A)

Why is de-emphasis commonly used in FM communications receivers?

A. For compatibility with transmitters using phase modulation

E7E07 (B)

What is meant by the term baseband in radio communications?

B. The frequency components present in the modulating signal

E7E08 (C)

What are the principal frequencies that appear at the output of a mixer circuit?

C. The two input frequencies along with their sum and difference frequencies

E7E09 (A)

What occurs when an excessive amount of signal energy reaches a mixer circuit?

A. Spurious mixer products are generated

E7E10 (A)

How does a diode detector function?

A. By rectification and filtering of RF signals

E7E11 (C)

Which of the following types of detector is well suited for demodulating SSB signals?

C. Product detector

E7E12 (D)

What is a frequency discriminator stage in a FM receiver?

D. A circuit for detecting FM signals

E7E13 (D)

Which of the following describes a common means of generating an SSB signal when using digital signal processing?

D. The quadrature method

E7E14 (C)

What is meant by direct conversion when referring to a software defined receiver?

C. Incoming RF is mixed to "baseband" for analog-to-digital conversion and subsequent processing

E7F Frequency markers and counters: frequency divider circuits; frequency marker generators; frequency counters

E7F01 (D)

What is the purpose of a prescaler circuit?

D. It divides a higher frequency signal so a low-frequency counter can display the input frequency

E7F02 (B)

Which of the following would be used to reduce a signal's frequency by a factor of ten?

B. A prescaler

E7F03 (A)

What is the function of a decade counter digital IC?

A. It produces one output pulse for every ten input pulses

E7F04 (C)

What additional circuitry must be added to a 100-kHz crystal-controlled marker generator so as to provide markers at 50 and 25 kHz?

C. Two flip-flops

E7F05 (D)

Which of the following is a technique for providing high stability oscillators needed for microwave transmission and reception?

A. Use a GPS signal reference

B. Use a rubidium stabilized reference oscillator

C. Use a temperature-controlled high Q dielectric resonator

D. All of these choices are correct

E7F06 (C)

What is one purpose of a marker generator?

C. To provide a means of calibrating a receiver's frequency settings

E7F07 (A)

What determines the accuracy of a frequency counter?

A. The accuracy of the time base

E7F08 (C)

Which of the following is performed by a frequency counter?

C. Counting the number of input pulses occurring within a specific period of time

E7F09 (A)

What is the purpose of a frequency counter?

A. To provide a digital representation of the frequency of a signal

E7F10 (B)

What alternate method of determining frequency, other than by directly counting input pulses, is used by some counters?

B. Period measurement plus mathematical computation

E7F11 (C)

What is an advantage of a period-measuring frequency counter over a direct-count type?

C. It provides improved resolution of low-frequency signals within a comparable time period

E7G Active filters and op-amps: active audio filters; characteristics; basic circuit design; operational amplifiers

E7G01 (B)

What primarily determines the gain and frequency characteristics of an op-amp RC active filter?

B. The values of capacitors and resistors external to the op-amp

E7G02 (D)

What is the effect of ringing in a filter?

D. Undesired oscillations added to the desired signal

E7G03 (D)

Which of the following is an advantage of using an op-amp instead of LC elements in an audio filter?

D. Op-amps exhibit gain rather than insertion loss

E7G04 (C)

Which of the following is a type of capacitor best suited for use in high-stability op-amp RC active filter circuits?

C. Polystyrene

E7G05 (A)

How can unwanted ringing and audio instability be prevented in a multi-section op-amp RC audio filter circuit?

A. Restrict both gain and Q

E7G06 (D)

Which of the following is the most appropriate use of an op-amp active filter?

D. As an audio filter in a receiver

E7G07 (C)

What magnitude of voltage gain can be expected from the circuit in Figure E7-4 when R1 is 10 ohms and RF is 470 ohms?

C. 47

E7G08 (D)

How does the gain of an ideal operational amplifier vary with frequency?

D. It does not vary with frequency

E7G09 (D)

What will be the output voltage of the circuit shown in Figure E7-4 if R1 is 1000 ohms, RF is 10,000 ohms, and 0.23 volts dc is applied to the input?

D. -2.3 volts

E7G10 (C)

What absolute voltage gain can be expected from the circuit in Figure E7-4 when R1 is 1800 ohms and RF is 68 kilohms?

C. 38

E7G11 (B)

What absolute voltage gain can be expected from the circuit in Figure E7-4 when R1 is 3300 ohms and RF is 47 kilohms?

B. 14

E7G12 (A)

What is an integrated circuit operational amplifier?

A. A high-gain, direct-coupled differential amplifier with very high input and very low output impedance

E7G13 (C)

What is meant by the term op-amp input-offset voltage?

C. The differential input voltage needed to bring the open-loop output voltage to zero

E7G14 (D)

What is the typical input impedance of an integrated circuit op-amp?

D. Very high

E7G15 (A)

What is the typical output impedance of an integrated circuit op-amp?

A. Very low

E7H Oscillators and signal sources: types of oscillators; synthesizers and phase-locked loops; direct digital synthesizers

E7H01 (D)

What are three oscillator circuits used in Amateur Radio equipment?

D. Colpitts, Hartley and Pierce

E7H02 (C)

What condition must exist for a circuit to oscillate?

C. It must have positive feedback with a gain greater than 1

E7H03 (A)

How is positive feedback supplied in a Hartley oscillator?

A. Through a tapped coil

E7H04 (C)

How is positive feedback supplied in a Colpitts oscillator?

C. Through a capacitive divider

E7H05 (D)

How is positive feedback supplied in a Pierce oscillator?

D. Through a quartz crystal

E7H06 (B)

Which of the following oscillator circuits are commonly used in VFOs?

B. Colpitts and Hartley

E7H07 (C)

What is a magnetron oscillator?

C. A UHF or microwave oscillator consisting of a diode vacuum tube with a specially shaped anode, surrounded by an external magnet

E7H08 (A)

What is a Gunn diode oscillator?

A. An oscillator based on the negative resistance properties of properly-doped semiconductors

E7H09 (A)

What type of frequency synthesizer circuit uses a phase accumulator, lookup table, digital to analog converter and a low-pass anti-alias filter?

A. A direct digital synthesizer

E7H10 (B)

What information is contained in the lookup table of a direct digital frequency synthesizer?

B. The amplitude values that represent a sine-wave output

E7H11 (C)

What are the major spectral impurity components of direct digital synthesizers?

C. Spurious signals at discrete frequencies

E7H12 (D)

Which of the following is a principal component of a direct digital synthesizer (DDS)?

D. Phase accumulator

E7H13 (A)

What is the capture range of a phase-locked loop circuit?

A. The frequency range over which the circuit can lock

E7H14 (C)

What is a phase-locked loop circuit?

C. An electronic servo loop consisting of a phase detector, a low-pass filter, a voltage-controlled oscillator, and a stable reference oscillator

E7H15 (D)

Which of these functions can be performed by a phase-locked loop?

D. Frequency synthesis, FM demodulation

E7H16 (B)

Why is the short-term stability of the reference oscillator important in the design of a phase locked loop (PLL) frequency synthesizer?

B. Any phase variations in the reference oscillator signal will produce phase noise in the synthesizer output

E7H17 (C)

Why is a phase-locked loop often used as part of a variable frequency synthesizer for receivers and transmitters?

C. It makes it possible for a VFO to have the same degree of frequency stability as a crystal oscillator

E7H18 (A)

What are the major spectral impurity components of phase-locked loop synthesizers?

A. Phase noise

SUBELEMENT E8 - SIGNALS AND EMISSIONS [4 Exam Questions - 4 Groups]

E8A AC waveforms: sine, square, sawtooth and irregular waveforms; AC measurements; average and PEP of RF signals; pulse and digital signal waveforms

E8A01 (A)

What type of wave is made up of a sine wave plus all of its odd harmonics?

A. A square wave

E8A02 (C)

What type of wave has a rise time significantly faster than its fall time (or vice versa)?

C. A sawtooth wave

E8A03 (A)

What type of wave is made up of sine waves of a given fundamental frequency plus all its harmonics?

A. A sawtooth wave

E8A04 (C)

What is equivalent to the root-mean-square value of an AC voltage?

C. The DC voltage causing the same amount of heating in a resistor as the corresponding RMS AC voltage

E8A05 (D)

What would be the most accurate way of measuring the RMS voltage of a complex waveform?

D. By measuring the heating effect in a known resistor

E8A06 (A)

What is the approximate ratio of PEP-to-average power in a typical single-sideband phone signal?

A. 2.5 to 1

E8A07 (B)

What determines the PEP-to-average power ratio of a single-sideband phone signal?

B. The characteristics of the modulating signal

E8A08 (A)

What is the period of a wave?

A. The time required to complete one cycle

E8A09 (C)

What type of waveform is produced by human speech?

C. Irregular

E8A10 (B)

Which of the following is a distinguishing characteristic of a pulse waveform?

B. Narrow bursts of energy separated by periods of no signal

E8A11 (D)

What is one use for a pulse modulated signal?

D. Digital data transmission

E8A12 (D)

What type of information can be conveyed using digital waveforms?

A. Human speech

B. Video signals

C. Data

D. All of these choices are correct

E8A13 (C)

What is an advantage of using digital signals instead of analog signals to convey the same information?

C. Digital signals can be regenerated multiple times without error

E8A14 (A)

Which of these methods is commonly used to convert analog signals to digital signals?

A. Sequential sampling

E8A15 (B)

What would the waveform of a stream of digital data bits look like on a conventional oscilloscope?

B. A series of pulses with varying patterns

E8B Modulation and demodulation: modulation methods; modulation index and deviation ratio; pulse modulation; frequency and time division multiplexing

E8B01 (D)

What is the term for the ratio between the frequency deviation of an RF carrier wave, and the modulating frequency of its corresponding FM-phone signal?

D. Modulation index

E8B02 (D)

How does the modulation index of a phase-modulated emission vary with RF carrier frequency (the modulated frequency)?

D. It does not depend on the RF carrier frequency

E8B03 (A)

What is the modulation index of an FM-phone signal having a maximum frequency deviation of 3000 Hz either side of the carrier frequency, when the modulating frequency is 1000 Hz?

A. 3

E8B04 (B)

What is the modulation index of an FM-phone signal having a maximum carrier deviation of plus or minus 6 kHz when modulated with a 2-kHz modulating frequency?

B. 3

E8B05 (D)

What is the deviation ratio of an FM-phone signal having a maximum frequency swing of plus-or-minus 5 kHz when the maximum modulation frequency is 3 kHz?

D. 1.67

E8B06 (A)

What is the deviation ratio of an FM-phone signal having a maximum frequency swing of plus or minus 7.5 kHz when the maximum modulation frequency is 3.5 kHz?

A. 2.14

E8B07 (A)

When using a pulse-width modulation system, why is the transmitter's peak power greater than its average power?

A. The signal duty cycle is less than 100%

E8B08 (D)

What parameter does the modulating signal vary in a pulse-position modulation system?

D. The time at which each pulse occurs

E8B09 (B)

What is meant by deviation ratio?

B. The ratio of the maximum carrier frequency deviation to the highest audio modulating frequency

E8B10 (C)

Which of these methods can be used to combine several separate analog information streams into a single analog radio frequency signal?

C. Frequency division multiplexing

E8B11 (B)

Which of the following describes frequency division multiplexing?

B. Two or more information streams are merged into a "baseband", which then modulates the transmitter

E8B12 (B)

What is digital time division multiplexing?

B. Two or more signals are arranged to share discrete time slots of a data transmission

E8C Digital signals: digital communications modes; CW; information rate vs. bandwidth; spread-spectrum communications; modulation methods

E8C01 (D)

Which one of the following digital codes consists of elements having unequal length?

D. Morse code

E8C02 (B)

What are some of the differences between the Baudot digital code and ASCII?

B. Baudot uses five data bits per character, ASCII uses seven or eight; Baudot uses two characters as shift codes, ASCII has no shift code

E8C03 (C)

What is one advantage of using the ASCII code for data communications?

C. It is possible to transmit both upper and lower case text

E8C04 (C)

What technique is used to minimize the bandwidth requirements of a PSK31 signal?

C. Use of sinusoidal data pulses

E8C05 (C)

What is the necessary bandwidth of a 13-WPM international Morse code transmission?

C. Approximately 52 Hz

E8C06 (C)

What is the necessary bandwidth of a 170-hertz shift, 300-baud ASCII transmission?

C. 0.5 kHz

E8C07 (A)

What is the necessary bandwidth of a 4800-Hz frequency shift, 9600-baud ASCII FM transmission?

A. 15.36 kHz

E8C08 (D)

What term describes a wide-bandwidth communications system in which the transmitted carrier frequency varies according to some predetermined sequence?

D. Spread-spectrum communication

E8C09 (A)

Which of these techniques causes a digital signal to appear as wide-band noise to a conventional receiver?

A. Spread-spectrum

E8C10 (A)

What spread-spectrum communications technique alters the center frequency of a conventional carrier many times per second in accordance with a pseudo-random list of channels?

A. Frequency hopping

E8C11 (B)

What spread-spectrum communications technique uses a high speed binary bit stream to shift the phase of an RF carrier?

B. Direct sequence

E8C12 (D)

What is the advantage of including a parity bit with an ASCII character stream?

D. Some types of errors can be detected

E8C13 (B)

What is one advantage of using JT-65 coding?

B. The ability to decode signals which have a very low signal to noise ratio

E8D Waveforms: measurement, peak-to-peak, RMS, average; Electromagnetic Waves: definition, characteristics, polarization

E8D01 (A)

Which of the following is the easiest voltage amplitude parameter to measure when viewing a pure sine wave signal on an analog oscilloscope?

A. Peak-to-peak voltage

E8D02 (B)

What is the relationship between the peak-to-peak voltage and the peak voltage amplitude of a symmetrical waveform?

B. 2:1

E8D03 (A)

What input-amplitude parameter is valuable in evaluating the signal-handling capability of a Class A amplifier?

A. Peak voltage

E8D04 (B)

What is the PEP output of a transmitter that develops a peak voltage of 30 volts into a 50-ohm load?

B. 9 watts

E8D05 (D)

If an RMS-reading AC voltmeter reads 65 volts on a sinusoidal waveform, what is the peak-to-peak voltage?

D. 184 volts

E8D06 (B)

What is the advantage of using a peak-reading wattmeter to monitor the output of a SSB phone transmitter?

B. It gives a more accurate display of the PEP output when modulation is present

E8D07 (C)

What is an electromagnetic wave?

C. A wave consisting of an electric field and a magnetic field oscillating at right angles to each other

E8D08 (D)

Which of the following best describes electromagnetic waves traveling in free space?

D. Changing electric and magnetic fields propagate the energy

E8D09 (B)

What is meant by circularly polarized electromagnetic waves?

B. Waves with a rotating electric field

E8D10 (D)

What type of meter should be used to monitor the output signal of a voice-modulated single-sideband transmitter to ensure you do not exceed the maximum allowable power?

D. A peak-reading wattmeter

E8D11 (A)

What is the average power dissipated by a 50-ohm resistive load during one complete RF cycle having a peak voltage of 35 volts?

A. 12.2 watts

E8D12 (D)

What is the peak voltage of a sinusoidal waveform if an RMS-reading voltmeter reads 34 volts?

D. 48 volts

E8D13 (B)

Which of the following is a typical value for the peak voltage at a standard U.S. household electrical outlet?

B. 170 volts

E8D14 (C)

Which of the following is a typical value for the peak-to-peak voltage at a standard U.S. household electrical outlet?

C. 340 volts

E8D15 (A)

Which of the following is a typical value for the RMS voltage at a standard U.S. household electrical power outlet?

A. 120V AC

E8D16 (A)

What is the RMS value of a 340-volt peak-to-peak pure sine wave?

A. 120V AC

SUBELEMENT E9 - ANTENNAS AND TRANSMISSION LINES [8 Exam Questions - 8 Groups]

E9A Isotropic and gain antennas: definitions; uses; radiation patterns; Basic antenna parameters: radiation resistance and reactance, gain, beamwidth, efficiency

E9A01 (C)

Which of the following describes an isotropic antenna?

C. A theoretical antenna used as a reference for antenna gain

E9A02 (B)

How much gain does a 1/2-wavelength dipole in free space have compared to an isotropic antenna?

B. 2.15 dB

E9A03 (D)

Which of the following antennas has no gain in any direction?

D. Isotropic antenna

E9A04 (A)

Why would one need to know the feed point impedance of an antenna?

A. To match impedances in order to minimize standing wave ratio on the transmission line

E9A05 (B)

Which of the following factors may affect the feed point impedance of an antenna?

B. Antenna height, conductor length/diameter ratio and location of nearby conductive objects

E9A06 (D)

What is included in the total resistance of an antenna system?

D. Radiation resistance plus ohmic resistance

E9A07 (C)

What is a folded dipole antenna?

C. A dipole constructed from one wavelength of wire forming a very thin loop

E9A08 (A)

What is meant by antenna gain?

A. The ratio relating the radiated signal strength of an antenna in the direction of maximum radiation to that of a reference antenna

E9A09 (B)

What is meant by antenna bandwidth?

B. The frequency range over which an antenna satisfies a performance requirement

E9A10 (B)

How is antenna efficiency calculated?

B. (radiation resistance / total resistance) x 100%

E9A11 (A)

Which of the following choices is a way to improve the efficiency of a ground-mounted quarter-wave vertical antenna?

A. Install a good radial system

E9A12 (C)

Which of the following factors determines ground losses for a ground-mounted vertical antenna operating in the 3-30 MHz range?

C. Soil conductivity

E9A13 (A)

How much gain does an antenna have compared to a 1/2-wavelength dipole when it has 6 dB gain over an isotropic antenna?

A. 3.85 dB

E9A14 (B)

How much gain does an antenna have compared to a 1/2-wavelength dipole when it has 12 dB gain over an isotropic antenna?

B. 9.85 dB

E9A15 (C)

What is meant by the radiation resistance of an antenna?

C. The value of a resistance that would dissipate the same amount of power as that radiated from an antenna

E9B Antenna patterns: E and H plane patterns; gain as a function of pattern; antenna design (computer modeling of antennas); Yagi antennas

E9B01 (B)

In the antenna radiation pattern shown in Figure E9-1, what is the 3-dB beamwidth?

B. 50 degrees

E9B02 (B)

In the antenna radiation pattern shown in Figure E9-1, what is the front-to-back ratio?

B. 18 dB

E9B03 (B)

In the antenna radiation pattern shown in Figure E9-1, what is the front-to-side ratio?

B. 14 dB

E9B04 (D)

What may occur when a directional antenna is operated at different frequencies within the band for which it was designed?

D. The gain may change depending on frequency

E9B05 (B)

What usually occurs if a Yagi antenna is designed solely for maximum forward gain?

B. The front-to-back ratio decreases

E9B06 (A)

If the boom of a Yagi antenna is lengthened and the elements are properly retuned, what usually occurs?

A. The gain increases

E9B07 (C)

How does the total amount of radiation emitted by a directional gain antenna compare with the total amount of radiation emitted from an isotropic antenna, assuming each is driven by the same amount of power?

C. They are the same

E9B08 (A)

How can the approximate beamwidth in a given plane of a directional antenna be determined?

A. Note the two points where the signal strength of the antenna is 3 dB less than maximum and compute the angular difference

E9B09 (B)

What type of computer program technique is commonly used for modeling antennas?

B. Method of Moments

E9B10 (A)

What is the principle of a Method of Moments analysis?

A. A wire is modeled as a series of segments, each having a uniform value of current

E9B11 (C)

What is a disadvantage of decreasing the number of wire segments in an antenna model below the guideline of 10 segments per half-wavelength?

C. The computed feed point impedance may be incorrect

E9B12 (D)

What is the far-field of an antenna?

D. The region where the shape of the antenna pattern is independent of distance

E9B13 (B)

What does the abbreviation NEC stand for when applied to antenna modeling programs?

B. Numerical Electromagnetics Code

E9B14 (D)

What type of information can be obtained by submitting the details of a proposed new antenna to a modeling program?

A. SWR vs. frequency charts

B. Polar plots of the far-field elevation and azimuth patterns

C. Antenna gain

D. All of these choices are correct

E9C Wire and phased vertical antennas: beverage antennas; rhombic antennas; elevation above real ground; ground effects as related to polarization; take-off angles

E9C01 (D)

What is the radiation pattern of two 1/4-wavelength vertical antennas spaced 1/2-wavelength apart and fed 180 degrees out of phase?

D. A figure-8 oriented along the axis of the array

E9C02 (A)

What is the radiation pattern of two 1/4-wavelength vertical antennas spaced 1/4-wavelength apart and fed 90 degrees out of phase?

A. A cardioid

E9C03 (C)

What is the radiation pattern of two 1/4-wavelength vertical antennas spaced 1/2-wavelength apart and fed in phase?

C. A Figure-8 broadside to the axis of the array

E9C04 (B)

Which of the following describes a basic unterminated rhombic antenna?

B. Bidirectional; four-sides, each side one or more wavelengths long; open at the end opposite the transmission line connection

E9C05 (C)

What are the disadvantages of a terminated rhombic antenna for the HF bands?

C. The antenna requires a large physical area and 4 separate supports

E9C06 (B)

What is the effect of a terminating resistor on a rhombic antenna?

B. It changes the radiation pattern from bidirectional to unidirectional

E9C07 (A)

What type of antenna pattern over real ground is shown in Figure E9-2?

A. Elevation

E9C08 (C)

What is the elevation angle of peak response in the antenna radiation pattern shown in Figure E9-2?

C. 7.5 degrees

E9C09 (B)

What is the front-to-back ratio of the radiation pattern shown in Figure E9-2?

B. 28 dB

E9C10 (A)

How many elevation lobes appear in the forward direction of the antenna radiation pattern shown in Figure E9-2?

A. 4

E9C11 (D)

How is the far-field elevation pattern of a vertically polarized antenna affected by being mounted over seawater versus rocky ground?

D. The low-angle radiation increases

E9C12 (D)

When constructing a Beverage antenna, which of the following factors should be included in the design to achieve good performance at the desired frequency?

D. It should be one or more wavelengths long

E9C13 (C)

What is the main effect of placing a vertical antenna over an imperfect ground?

C. It reduces low-angle radiation

E9D Directional antennas: gain; satellite antennas; antenna beamwidth; stacking antennas; antenna efficiency; traps; folded dipoles; shortened and mobile antennas; grounding

E9D01 (C)

How does the gain of an ideal parabolic dish antenna change when the operating frequency is doubled?

C. Gain increases by 6 dB

E9D02 (C)

How can linearly polarized Yagi antennas be used to produce circular polarization?

C. Arrange two Yagis perpendicular to each other with the driven elements at the same point on the boom and fed 90 degrees out of phase

E9D03 (D)

How does the beamwidth of an antenna vary as the gain is increased?

D. It decreases

E9D04 (A)

Why is it desirable for a ground-mounted satellite communications antenna system to be able to move in both azimuth and elevation?

A. In order to track the satellite as it orbits the Earth

E9D05 (A)

Where should a high-Q loading coil be placed to minimize losses in a shortened vertical antenna?

A. Near the center of the vertical radiator

E9D06 (C)

Why should an HF mobile antenna loading coil have a high ratio of reactance to resistance?

C. To minimize losses

E9D07 (A)

What is a disadvantage of using a multiband trapped antenna?

A. It might radiate harmonics

E9D08 (B)

What happens to the bandwidth of an antenna as it is shortened through the use of loading coils?

B. It is decreased

E9D09 (D)

What is an advantage of using top loading in a shortened HF vertical antenna?

D. Improved radiation efficiency

E9D10 (A)

What is the approximate feed point impedance at the center of a two-wire folded dipole antenna?

A. 300 ohms

E9D11 (D)

What is the function of a loading coil as used with an HF mobile antenna?

D. To cancel capacitive reactance

E9D12 (D)

What is one advantage of using a trapped antenna?

D. It may be used for multiband operation

E9D13 (B)

What happens to feed point impedance at the base of a fixed-length HF mobile antenna as the frequency of operation is lowered?

B. The radiation resistance decreases and the capacitive reactance increases

E9D14 (B)

Which of the following types of conductor would be best for minimizing losses in a station's RF ground system?

B. A wide flat copper strap

E9D15 (C)

Which of the following would provide the best RF ground for your station?

C. An electrically-short connection to 3 or 4 interconnected ground rods driven into the Earth

E9E Matching: matching antennas to feed lines; power dividers

E9E01 (B)

What system matches a high-impedance transmission line to a lower impedance antenna by connecting the line to the driven element in two places spaced a fraction of a wavelength each side of element center?

B. The delta matching system

E9E02 (A)

What is the name of an antenna matching system that matches an unbalanced feed line to an antenna by feeding the driven element both at the center of the element and at a fraction of a wavelength to one side of center?

A. The gamma match

E9E03 (D)

What is the name of the matching system that uses a section of transmission line connected in parallel with the feed line at or near the feed point?

D. The stub match

E9E04 (B)

What is the purpose of the series capacitor in a gamma-type antenna matching network?

B. To cancel the inductive reactance of the matching network

E9E05 (A)

How must the driven element in a 3-element Yagi be tuned to use a hairpin matching system?

A. The driven element reactance must be capacitive

E9E06 (C)

What is the equivalent lumped-constant network for a hairpin matching system on a

3-element Yagi?

C. L network

E9E07 (B)

What term best describes the interactions at the load end of a mismatched transmission line?

B. Reflection coefficient

E9E08 (D)

Which of the following measurements is characteristic of a mismatched transmission line?

D. An SWR greater than 1:1

E9E09 (C)

Which of these matching systems is an effective method of connecting a 50-ohm coaxial cable feed line to a grounded tower so it can be used as a vertical antenna?

C. Gamma match

E9E10 (C)

Which of these choices is an effective way to match an antenna with a 100-ohm feed point impedance to a 50-ohm coaxial cable feed line?

C. Insert a 1/4-wavelength piece of 75-ohm coaxial cable transmission line in series between the antenna terminals and the 50-ohm feed cable

E9E11 (B)

What is an effective way of matching a feed line to a VHF or UHF antenna when the impedances of both the antenna and feed line are unknown?

B. Use the universal stub matching technique

E9E12 (A)

What is the primary purpose of a phasing line when used with an antenna having multiple driven elements?

A. It ensures that each driven element operates in concert with the others to create the desired antenna pattern

E9E13 (C)

What is the purpose of a Wilkinson divider?

C. It divides power equally among multiple loads while preventing changes in one load from disturbing power flow to the others

E9F Transmission lines: characteristics of open and shorted feed lines; 1/8 wavelength; 1/4 wavelength; 1/2 wavelength; feed lines: coax versus open-wire; velocity factor; electrical length; coaxial cable dielectrics; velocity factor

E9F01 (D)

What is the velocity factor of a transmission line?

D. The velocity of the wave in the transmission line divided by the velocity of light in a vacuum

E9F02 (C)

Which of the following determines the velocity factor of a transmission line?

C. Dielectric materials used in the line

E9F03 (D)

Why is the physical length of a coaxial cable transmission line shorter than its electrical length?

D. Electrical signals move more slowly in a coaxial cable than in air

E9F04 (B)

What is the typical velocity factor for a coaxial cable with solid polyethylene dielectric?

B. 0.66

E9F05 (C)

What is the approximate physical length of a solid polyethylene dielectric coaxial transmission line that is electrically one-quarter wavelength long at 14.1 MHz?

C. 3.5 meters

E9F06 (C)

What is the approximate physical length of an air-insulated, parallel conductor transmission line that is electrically one-half wavelength long at 14.10 MHz?

C. 10 meters

E9F07 (A)

How does ladder line compare to small-diameter coaxial cable such as RG-58 at 50 MHz?

A. Lower loss

E9F08 (A)

What is the term for the ratio of the actual speed at which a signal travels through a transmission line to the speed of light in a vacuum?

A. Velocity factor

E9F09 (B)

What is the approximate physical length of a solid polyethylene dielectric coaxial transmission line that is electrically one-quarter wavelength long at 7.2 MHz?

B. 6.9 meters

E9F10 (C)

What impedance does a 1/8-wavelength transmission line present to a generator when the line is shorted at the far end?

C. An inductive reactance

E9F11 (C)

What impedance does a 1/8-wavelength transmission line present to a generator when the line is open at the far end?

C. A capacitive reactance

E9F12 (D)

What impedance does a 1/4-wavelength transmission line present to a generator when the line is open at the far end?

D. Very low impedance

E9F13 (A)

What impedance does a 1/4-wavelength transmission line present to a generator when the line is shorted at the far end?

A. Very high impedance

E9F14 (B)

What impedance does a 1/2-wavelength transmission line present to a generator when the line is shorted at the far end?

B. Very low impedance

E9F15 (A)

What impedance does a 1/2-wavelength transmission line present to a generator when the line is open at the far end?

A. Very high impedance

E9F16 (D)

Which of the following is a significant difference between foam-dielectric coaxial cable and solid-dielectric cable, assuming all other parameters are the same?

A. Reduced safe operating voltage limits

B. Reduced losses per unit of length

C. High velocity factor

D. All of these choices are correct

E9G The Smith chart

E9G01 (A)

Which of the following can be calculated using a Smith chart?

A. Impedance along transmission lines

E9G02 (B)

What type of coordinate system is used in a Smith chart?

B. Resistance circles and reactance arcs

E9G03 (C)

Which of the following is often determined using a Smith chart?

C. Impedance and SWR values in transmission lines

E9G04 (C)

What are the two families of circles and arcs that make up a Smith chart?

C. Resistance and reactance

E9G05 (A)

What type of chart is shown in Figure E9-3?

A. Smith chart

E9G06 (B)

On the Smith chart shown in Figure E9-3, what is the name for the large outer circle on which the reactance arcs terminate?

B. Reactance axis

E9G07 (D)

On the Smith chart shown in Figure E9-3, what is the only straight line shown?

D. The resistance axis

E9G08 (C)

What is the process of normalization with regard to a Smith chart?

C. Reassigning impedance values with regard to the prime center

E9G09 (A)

What third family of circles is often added to a Smith chart during the process of solving problems?

A. Standing-wave ratio circles

E9G10 (D)

What do the arcs on a Smith chart represent?

D. Points with constant reactance

E9G11 (B)

How are the wavelength scales on a Smith chart calibrated?

B. In fractions of transmission line electrical wavelength

E9H Effective radiated power; system gains and losses; radio direction finding antennas

E9H01 (D)

What is the effective radiated power relative to a dipole of a repeater station with 150 watts transmitter power output, 2-dB feed line loss, 2.2-dB duplexer loss and 7-dBd antenna gain?

D. 286 watts

E9H02 (A)

What is the effective radiated power relative to a dipole of a repeater station with 200 watts transmitter power output, 4-dB feed line loss, 3.2-dB duplexer loss, 0.8-dB circulator loss and 10-dBd antenna gain?

A. 317 watts

E9H03 (B)

What is the effective isotropic radiated power of a repeater station with 200 watts transmitter power output, 2-dB feed line loss, 2.8-dB duplexer loss, 1.2-dB circulator loss and 7-dBi antenna gain?

B. 252 watts

E9H04 (C)

What term describes station output, including the transmitter, antenna and everything in between, when considering transmitter power and system gains and losses?

C. Effective radiated power

E9H05 (A)

What is the main drawback of a wire-loop antenna for direction finding?

A. It has a bidirectional pattern

E9H06 (C)

What is the triangulation method of direction finding?

C. Antenna headings from several different receiving locations are used to locate the signal source

E9H07 (D)

Why is it advisable to use an RF attenuator on a receiver being used for direction finding?

D. It prevents receiver overload which could make it difficult to determine peaks or nulls

E9H08 (A)

What is the function of a sense antenna?

A. It modifies the pattern of a DF antenna array to provide a null in one direction

E9H09 (C)

Which of the following describes the construction of a receiving loop antenna?

C. One or more turns of wire wound in the shape of a large open coil

E9H10 (D)

How can the output voltage of a multi-turn receiving loop antenna be increased?

D. By increasing either the number of wire turns in the loop or the area of the loop structure or both

E9H11 (B)

What characteristic of a cardioid-pattern antenna is useful for direction finding?

B. A very sharp single null

E9H12 (B)

What is an advantage of using a shielded loop antenna for direction finding?

B. It is electro-statically balanced against ground, giving better nulls

SUBELEMENT E0 – SAFETY - [1 exam question — 1 group]

E0A Safety: amateur radio safety practices; RF radiation hazards; hazardous materials

E0A01 (C)

What, if any, are the differences between the radiation produced by radioactive materials and the electromagnetic energy radiated by an antenna?

C. Radioactive materials emit ionizing radiation, while RF signals have less energy and can only cause heating

E0A02 (B)

When evaluating RF exposure levels from your station at a neighbor's home, what must you do?

B. Make sure signals from your station are less than the uncontrolled MPE limits

E0A03 (C)

Which of the following would be a practical way to estimate whether the RF fields produced by an amateur radio station are within permissible MPE limits?

C. Use an antenna modeling program to calculate field strength at accessible locations

E0A04 (C)

When evaluating a site with multiple transmitters operating at the same time, the operators and licensees of which transmitters are responsible for mitigating over-exposure situations?

C. Each transmitter that produces 5% or more of its MPE exposure limit at accessible locations

E0A05 (B)

What is one of the potential hazards of using microwaves in the amateur radio bands?

B. The high gain antennas commonly used can result in high exposure levels

E0A06 (D)

Why are there separate electric (E) and magnetic (H) field MPE limits?

A. The body reacts to electromagnetic radiation from both the E and H fields

B. Ground reflections and scattering make the field impedance vary with location

C. E field and H field radiation intensity peaks can occur at different locations

D. All of these choices are correct

E0A07 (B)

How may dangerous levels of carbon monoxide from an emergency generator be detected?

B. Only with a carbon monoxide detector

E0A08 (C)

What does SAR measure?

C. The rate at which RF energy is absorbed by the body

E0A09 (C)

Which insulating material commonly used as a thermal conductor for some types of electronic devices is extremely toxic if broken or crushed and the particles are accidentally inhaled?

C. Beryllium Oxide

E0A10 (A)

What material found in some electronic components such as high-voltage capacitors and transformers is considered toxic?

A. Polychlorinated biphenyls

E0A11 (C)

Which of the following injuries can result from using high-power UHF or microwave transmitters?

C. Localized heating of the body from RF exposure in excess of the MPE limits

The Complete Study Guide For All Amateur Radio Tests

~~end of question pool text~~

NOTE: The graphics required for certain questions in sections E5, E6, E7, and E9 are included on the following pages:

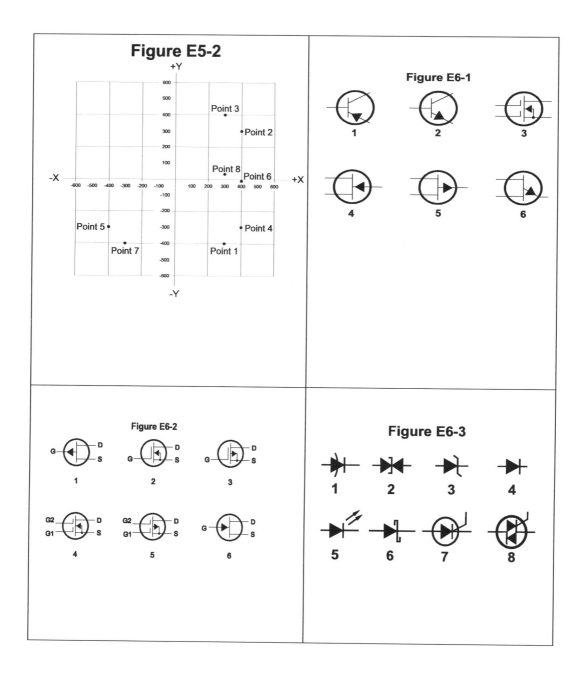

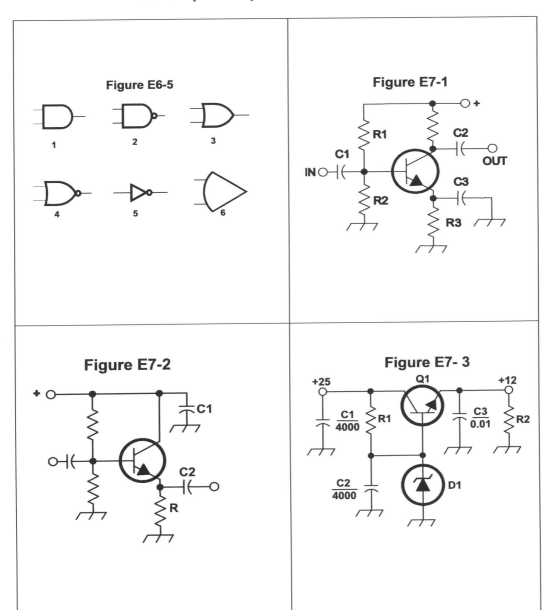

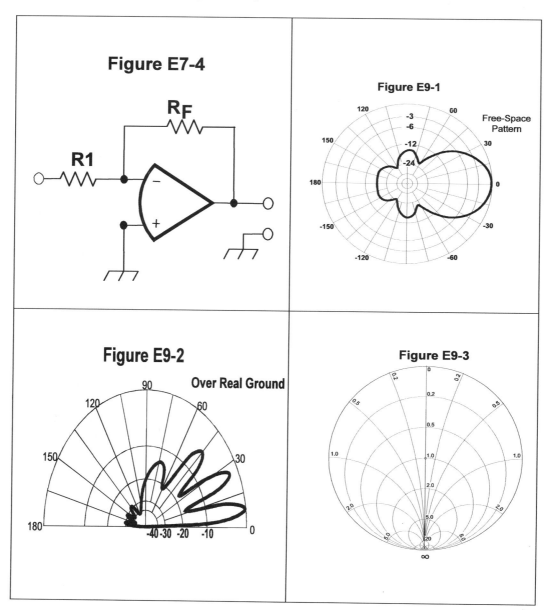

~~~ end of question pool ~~~

Volunteer Examiner
Your chance to give back to the Ham Radio Community.

Prior to 1982, examinations for amateur radio licenses were conducted by the FCC directly. Candidates had to travel to an FCC office at times set by the FCC. In some cases it was possible to obtain an introductory or novice license through a field examination but at one time these were only good for a year and could not be renewed. With the election of Ronald Reagan came cost-cutting measures throughout the entire federal government. Amateur radio program came under fire and could have been closed because the government could no longer afford the bill for conducting examinations of amateur radio licensees.

Senator Barry Goldwater (formerly K7UGA), an avid ham, was an influential member of Congress at this time He authored and pushed through the law that established the Volunteer Examination (VE) system for amateur radio licensing, in so doing sparing amateur radio from what would otherwise been almost certain death at the hands of Reagan's "pay-as-you-go" funding approach for federal agencies. From the FCC's standpoint, the VE system took nearly all the expense of the testing system away from the FCC. Exams would be conducted by volunteer teams organized and supervised by FCC-approved Volunteer Examiner Coordinators (VECs). The examinations to be used would also be developed by the coordinators, subject to approval by the FCC. The FCC's expense was reduced to the very minimal effort of supervising the VECs. The Goldwater bill also extended the normal license term to ten years and made all licenses renewable. Testing procedures moved to the local county or community. This made it easier for most hams. Our numbers grew.

Each volunteer examiner team consists of at least three Volunteer Examiners (VEs) each accredited by a Volunteer Examiner Coordinator (VEC). To be a volunteer examiner, an individual must be a currently licensed amateur radio operator at least 18 years of age whose license has never been suspended or revoked. An individual may be accredited by more than one VEC (many VEs are accredited by multiple VECs, although personally I'm only accredited with the ARRL/VEC. Go to www.arrl.org for more information. All VEs at a given session must be accredited by the coordinator that is sponsoring that session.

The VEs administering an exam must be of a license class greater than the license that the exam being administered would qualify the candidate for, except that an Extra VE may administer any element (since there is no license class greater than Extra).

The test to become a VE (at least as managed by the ARRL/VEC) is not very difficult, and it's not uncommon for Extras to decide to take this "next step" in the service to the hobby.

ARRL VE Accreditation Material

Open Book Review and Exam

This document contains the following information, which has been extracted from the ARRL VE manual:

1. When may all or a portion of the test fees collected by a VE team be used to offset expenses not related to examinations (such as instruction or club expenses)?

Never

2. By how much can a VEC or a VE team vary the test fee that it charges each candidate to offset expenses incurred in coordinating sessions during the calendar year?

None / $0.00

3. Which FCC bureau administers the VE Program?

FCC Wireless Telecommunications Bureau

4. Who is responsible for maintaining question pools from which all written element exams must be designed and assembled?

National Conference of VECs, Question Pool Committee

5. How old must an amateur be to be eligible for VE accreditation?

18

6. May an amateur be accredited by a VEC if his/her license was ever suspended or revoked? **NO**

7. May a VE be accredited concurrently by more than one VEC?

YES

8. What is the name of the ARRL/VEC program feature where an ARRL/VEC VE team (with the appropriate documentation provided) can accept the services of a VE who is accredited with a different VEC?

instant accreditation

9. How long is the term of accreditation for ARRL/VEC Volunteer Examiners?

concurrent with license term

10. What is the minimum number of VEs needed to conduct a VEC coordinated test session?

3

11. Who must grade the examinations of each examinee?

Each administering VE

12. May VEs who are related, e.g. husband/wife, administer elements at the same session?

YES

May they administer exams to their relatives?

NO

13. What License application form is used by an examinee to apply for an amateur operator license at a test session?

NVEC 605

14. Which element(s) can be administered by a General class VE? __**2 – technician**__
Which element(s) can be administered by an Advanced class VE? ___**2,3 – technician, general**___

15. In the ARRL/VEC program may a VE Team Liaison (or VE Session Manager) who holds a General or Advanced class license receive test papers for Elements 2, 3 and 4?

NO

16. Which license class, if any, must the VE Team Contact Person hold?

no license required

17. Who decides when and where a test session will be held?

VE team (Volunteer Examiner team)

18. What is the test fee charged to each candidate who takes examination elements for Technician or higher class licenses, including upgrades, at an ARRL/VEC coordinated session during this calendar year?

$15

19. Under what circumstances may a VE team conduct a test session without having publicly announced it in advance?

It is no longer required by law, but ARRL policy says You Shall announce everything, but then the manual mentions giving tests at bedside if required.... Publicizing a willingness to do special sessions to accommodate individual needs is good, but publishing an individual special session seems to single out the person you are trying to help

20. If tests are given at a convention or hamfest, under what conditions may an examinee be required to pay to gain access to the test site?

only if they will be participating in the event (hamfest) also

21. How long does an average ARRL/VEC test session last?

3-1/2 hours

22. What VEC form is used for recording each examinee's progress and pass/fail results for a test session?

Test Session Report / Candidate Roster (both seem to fit that criteria, but Candidate Roster seems more correct)

23. Which application form is the only form currently accepted by the FCC for amateur license renewal or address change requests filed directly with FCC by mail or on line?

FCC Form 605

24. Under what conditions can Form 605 applications be privately reproduced?

To meet immediate needs

25. If an unlicensed applicant passes Elements 2 and 3 at an ARRL/VEC coordinated session, what new license class (earned) should be indicated on the CSCE and on the Form 605?

General

26. If an unlicensed applicant passes Elements 2, 3 and 4 at an ARRL/VEC coordinated session, what new license class (earned) should be indicated on the CSCE and Form 605?

Extra

27. Can applicants who are seeking an address change send an "NCVEC Form 605" directly to the FCC?

NO

28. Can a VE, who has recently upgraded, serve as a VE exercising the privileges of the new higher class license before the newly upgraded license grant appears in the FCC's Amateur Service license data base?

NO

29. In the ARRL/VEC program, within ten days after a test session is administered, where must all NCVEC Form 605 applications for successful applicants and all exam documents be sent (this includes any test documents passed or failed, or any other documents written on by the examinee) ?

__Coordinating VEC__

30. May VE Teams who are not officially field stocked by ARRL/VEC retain test booklets or exam materials after the conclusion of the test session?

NO

31. May an applicant who is applying for an upgrade at a test session also request a Vanity call using that same upgrade application (NCVEC Form 605)?

NO

32. Will persons served at a VEC coordinated examination, who submit an application for an upgrade, address change or a systematic call sign change, receive a freshly renewed 10 year term license from the FCC if their license is not to expire in 90 days or less?

NO

33. Under current FCC Rules/procedures, can an amateur license be renewed before 90 days? **NO**

34. In the ARRL/VEC program, what credit can be issued an applicant who successfully completes one or more elements, but who cannot document successful completion of lower elements?

Element Only Credit

35. For how long is credit, as shown on a CSCE, valid?

365 days

36. May the VE Team deny an applicant the use of a calculator? If so, when?

Yes – if it cannot be shown to be free of stored values, formulas or programs

37. What must a Technician class amateur, who passed the technician exam before 3/21/1987, do to receive credit for Element 3 towards a General class or higher upgrade?

Show previous license, original CSCE, or other proof such as inclusion in a published call sign lookup database from that timeframe

38. Should elements that an applicant successfully completed at an earlier (different test date/location) test session also be indicated on a CSCE issued at your session today (current day) or at the next test session (future date) where the applicant successfully completes an additional element?

NO

39. To whom are the three copies of the ARRL/VEC CSCE to be distributed?

White/Blue- Candidate, Pink – ARRL, Yellow – VE Team

40. According to ARRL/VEC policy, how soon should a VE team mail applications for successful (upgrading) candidates to the coordinating VEC after the test session has been conducted?

as soon as possible – 10 days max

The Complete Study Guide For All Amateur Radio Tests

General Hints

Have everything you need to take the test when you walk into the test session. Lay it out before you leave. Check the list twice before leaving the house.

1. A valid photo ID (like your driver's license or a passport).

2. Two sharpened pencils and one pen. You should take your tests with a pencil in case you wish to correct an answer but always sign your name with a pen.

3. A calculator (If you use this manual you will likely not need one but if you wish to work out the formulas a calculator may be needed. It can not be programmable or have anything kept in memory. This will NOT be permitted.

4. If you are upgrading you need to bring the original **and** a copy of your current Amateur Radio license.

5. Examination fee - $15.00.

When you pass your test and there is enough time left you can take multiple tests in a single session. I have known people to go all the way from Technician to Extra Class in a single sitting, but do not force yourself. Do only what you are comfortable with.

The VEs are only required to tell you if you passed or failed. Some may offer more information but they cannot tell you about how you did on specific questions or topics.

If you fail and wish to retake a test you must pay another examination fee.

If you wish to take more than one test there will be a $15.00 fee per test.

Be sure to arrive ON TIME. The VE team can refuse to test you if you arrive after testing has begun.

Conduct yourself as if you were in a library – very quiet – Others are testing.

Hints for Taking the Written Exams

The Technician, General, and Amateur Extra exams are all multiple choice exams chosen from the "question pool" for that exam. This manual has exposed you to ALL possible questions in each pool. You will recognize the answers. You have seen them dozens of times before.

The Complete Study Guide For All Amateur Radio Tests

The Technician and General exams are 35 questions each of which you must get 26 correct to pass.

The Amateur Extra exam is 50 questions of which you must get 37 correct to pass the test.

Each of the questions has a "pool" of pre-defined questions that are selected at random for each of the actual test questions. The questions are presented in a random order and the answers are also randomized. You can't memorize the fact that question T1A04 is answer C and expect to get it right. You have to know the actual answers.

The question pools are periodically changed and updated.

Since this is a multiple choice test some simple tips for multiple choice tests are in order:

- Read ALL of the answers before you pick one. Many times the answers can sound similar.
- If you don't know the answer, skip to the next question and come back to it later. When you come back you may find that your subconscious mind has been working on the question and already has the answer ready for you.
- Use the process of elimination - some answers are obviously wrong so concentrate on the ones that are left. In a multiple choice test you can usually eliminate two answers since they will obviously be incorrect. Now you have a 50/50 chance to recognize an answer you have seen many times.
- Make sure the answer you mark on the answer sheet is for the question you just read - (put your answer for question 23 on answer sheet number 23). Mark the answer clearly.
- You have plenty of time - go over the questions again but only change answers if you are absolutely sure you made a mistake.

You can do this!

Good Luck!

Joseph Lumpkin
AB4AN

Theory and Formulas

ELECTRONICS and ANTENNAS

The most basic and meaningful law is this: Power in must equal power out. Energy can never be destroyed. It can only be emitted (conducted), reflected, or absorbed and turned into heat. This is true from resistors to antennas. Energy can never be lost.

In the mathematical formulas of electronic theory there are symbols used to represent various forces. E stands for voltage, I stands for current, R stands for resistance, P stands for power, and so on.

Voltage, (E) is the energy that pushes electrons through a conductor, such as a wire. Electrons are negatively charged so when a difference of potential energy is seen in the wire the electrons flow by being pushed away from the negative side and drawn to the positive side. This is because like charges repel and opposites attract. This explains many relationships also. The higher the voltage the greater the force pushing the electrons. This force is called "VOLTAGE" and is measured in units of VOLTS. Look at voltage like water pressure. It is the force pushing the water through the pipes like voltage pushes electrons through the wires.

There are items, such as a resistor, that have the ability to reduce or dissipate voltage. They do this by offering a path in which the electrons do not move as easily. It is the substance making up resistors that inhibit electron flow. Energy is lost through heat, since energy can never be destroyed. The difference in voltage between the points on each side of the resistor is known as the voltage drop. It represents a difference in potential energy.

Voltage can be delivered in a steady state, as with a battery, or it can be delivered in cycles, called Hertz, such as the 60-hertz voltage coming from the wall socket of our homes. When voltage is constant it is said to be Direct Current or D.C. When voltage alternates in cycles it is said to be Alternating Current or A.C.

Current, (I) is the amount of electrons flowing and is measured in Amperes or Amps. As voltage is the pressure of flow so current is the amount of flow. You can have a lot of pressure with few electrons. That would be like a fast flowing little stream. You can have a lot of water but not much flow. This would be like a pond. To do work you need both a lot of electrons (current) and a lot of pressure (voltage) The unit of work produced with both current and pressure ("I" and "E") is Power (P), measured in Watts. Therefore $P = I \times E$

Whether AC or DC current only flows through a circuit when a voltage source is connected to it with its "flow" being limited to both the resistance of the circuit and the voltage source pushing it. Also, as AC currents (and voltages) are periodic and vary with time the "effective" or "RMS", (Root Mean Squared) produces the same average power loss equivalent to a DC current. In other words, in an A.C. circuit there are parts of the wave that cannot be used effectively as it is changing directions. To get the usable voltage you multiply the peak voltage by .707. Peak times .707 = RMS. RMS x 1.414 = peak. Another way of measuring the usable energy is to place a resistor of known value into the circuit and measure the heat radiated by the resistor. The resistor is seeing the voltage and resisting it. The energy is being converted to heat because… say it with me now… Energy is never lost. It must go somewhere so it turns to heat.

Resistance

The **Resistance**, (R) of a circuit or component is its ability to prevent the flow of electrons. Resistance is measured in **Ohms**, and the Greek symbol Omega. If a circuit passes no electrons the resistance measures infinity. If it offers no resistance it measures zero resistance. Zero resistance is a short circuit. Infinite resistance is an open circuit.

The resistance measurement in Ohms is converted to a color code and painted on the resistor.

First the code

Black brown red orange yellow green blue violet Gray white
0   1   2   3   4   5   6   7   8   9

The mnemonic

Bad Boys Ravish our Young Girls But Violet Gives Willingly.

Color	Significant figures	Multiplier	Tolerance	Temp. Coefficient (ppm/K)	
Black	0	$\times 10^0$	–	250	U
Brown	1	$\times 10^1$	±1% F	100	S
Red	2	$\times 10^2$	±2% G	50	R

Color	Digit	Multiplier	Tolerance			
Orange	3	$\times 10^3$	–		15	P
Yellow	4	$\times 10^4$	($\pm 5\%$)	–	25	Q
Green	5	$\times 10^5$	$\pm 0.5\%$	D	20	Z
Blue	6	$\times 10^6$	$\pm 0.25\%$	C	10	Z
Violet	7	$\times 10^7$	$\pm 0.1\%$	B	5	M
Gray	8	$\times 10^8$	$\pm 0.05\%$ ($\pm 10\%$)	A	1	K
White	9	$\times 10^9$	–		–	
Gold	–	$\times 10^{-1}$	$\pm 5\%$	J		–
Silver	–	$\times 10^{-2}$	$\pm 10\%$	K		–
None	–	–	$\pm 20\%$	M		–

How to read the code

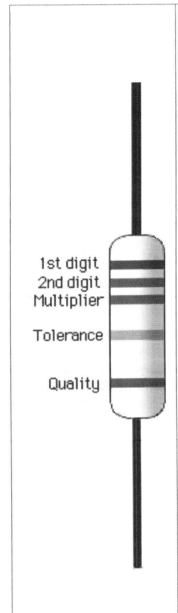

Start from the band closest to the end of the resistor. On a typical resistor there will be more space on one end than the other to the first band. If they are about even look for a tolerance band of gold (5%) or silver (10%) and start from the other side of the resistor from those bands.

There will be three bands or three bands and a tolerance band. The first two will be read as the number itself. Brown is one, red is two, orange is three, and so on. The last band tells how many zeros to add. Brown means place one zero at the end, red means two zeros, orange means add three zeros, and so on down the list.

For example, if a resistor had bands of green, red, and brown the resistance would be 520 ohms because green = 5, red = 2, and brown = 1 but since brown is the last colored band it says how many zeros to add so we add one zero to 52. The last band is known as the multiplier.

The next band is the tolerance band. It says the resistor will have no more variance than a certain percent. Gold is 5%, silver is 10% and no band is 20% variance.

Occasionally there will be a band apart from the others and at the extreme opposite end. In over 20 years of electronics repair I can count on one hand the number of times I have seen the fifth band. It is known as the quality band. Read the number as the '% Failure rate per 1000 hour' This is rated assuming full wattage being applied to the resistors. (To get better failure rates, resistors are typically specified to have twice the needed wattage dissipation that the circuit produces). Some resistors use this band for temperature coefficient information. This is how much the resistance will vary as the temperature of the component climbs.

Low resistance or high conduction means the material is a good conductor of electricity, such as copper, aluminum, or iron. High resistance means the material is an insulator such as plastic, glass, porcelain, clay, dry wood or paper.

The standard unit of measurement given for conductance is the **Siemen**, symbol (**S**). Conductivity may also be measured as the reciprocal of resistance. This unit is called a Mho (mo) and is the opposite of an Ohm.

The size of wire is important when it comes to conduction. The diameter or the length of the wire is like that of a pipe. The bigger the pipe the more water can flows through it. The smaller the diameter of the pipe the larger the resistance will be to the flow. When you try to push too much current through a small wire there will be resistance and the wire will heat up. There are actually wire round resistors that use this principal.

When electricity is passed through a resistor the voltage drops. Since energy can never be lost it turns to heat. To much heat can alter the makeup of the resistor and change its value. The amount of energy across the resistor can be calculated by measuring heat from the resistor. This is one way of measuring the RMS value in an A.C. circuit.

When it comes to insulators we want no current flow at all. Usually, with such insulators the more of an insulator the greater the voltage that can be present across the insulation without arcing. But heat can change that. As things carbonize they begin to conduct. Carbon is a conductor. Some resistors mix clay, which is an insulator, with carbon, a conductor, to form resistors with various resistance values. In an A.C. circuit, such as with Amateur Radio, frequency affects impedance, which is resistance in an A.C. circuit due to inductance or capacitance. The type and size of insulator inside a feed line like a co-axial cable will produce various impedance values according to what frequency is being used. Most radios are built to transmit and receive into a 50-52 ohm line.

Resistance, such as a clay-carbon resistor present the same resistance in an A.C. or D.C. circuit, regardless of frequency. There is a type of resistance called "Impedance" or Reactance" that is affected greatly by frequency. Reactance is seen in coils and capacitors implemented in an A.C. circuit

There are components that are neither resistors nor conductors. They are called "Semiconductors." They conduct well only under certain conditions. They are made up of silicon, germanium or some such material with measured amounts of impurities that change the conductions values. These materials have an imbalance of electrons so when a small charge is applied they begin to conduct. Semiconductors are used in the manufacturing of diodes, transistors and integrated circuits.

There is a relationship between Voltage (E), Current (I), and Resistance (R).

The relationship is summed up in Ohm's Law. **Ohm's Law** is defined as: $E = I \times R$.

Simply speaking, there is a relationship between Voltage, Current, and Resistance (or Reactance, as it is expressed in an A.C. circuit.) D.C. standing for Direct Current. It is a constant flow of electrons. A.C. stands for Alternating Current. In an A.C. circuit electron flow changes direction. The amount of times this change happens and comes back to its original direction is a cycle or 1 Hertz. The relationship between Voltage (symbol "E"), Current (symbol "I"), and Resistance (symbol "R") is expressed in Ohm's Law. R - the resistance in Ohms is written with the Greek symbol Ω.

There is also a relationship between Power, expressed in Watts (symbol "P"), Voltage (E) and Current (I). This relationship is expressed in Watts Law. E=IR.

One way to understand basic electronic theory is to compare electricity to the flow of water. Water needs both force and volume to do work. One could have a large amount of pressure but not a lot of volume. This would be like a strong stream from a small hose. There could also be a large volume of water with no pressure, like a still lake. This could not produce work either. But if there is a large volume with pressure to get from one place to another there is energy that can be harnessed to do work. Voltage (E) is pressure. Current (I) is the amount or volume. Watts (P) is power. This is the power to do work. P=IE.

If something resists the flow, such as a resistor in an A.C. or D.C. circuit or a capacitor or inductor in an A.C. circuit, the resistance or reactance will affect the power, voltage, and current of the circuit.

OHM's LAW

Voltage = E or V, Current = I, Power = P, Resistance = R, Reactance = Z

L = Inductance, C= Capacitance, F=Frequency

Voltage is measured in Volts, Current is measured in Amperes or Amps, Power is measured in Watts, Resistance and Reactance is measured in Ohms, Capacitance is measured in Farads, Inductance is measured in Henrys, Frequency is measured in Hertz or Cycles per second.

For simple voltage, current and resistive DC loads

To find current E/R.   I=E/R

To find resistance E/I.   R=E/I

To find voltage I*R. E=IR

For simple power calculations of DC circuits

To find power E*I. P=EI
To find power E^2/I. P= E^2/I
To find resistance E^2/I. R= E^2/I
To find power I^2*R. P= I^2*R.

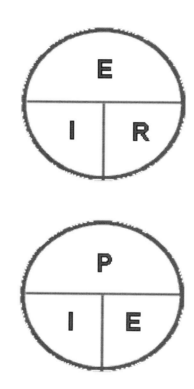

The Complete Study Guide For All Amateur Radio Tests

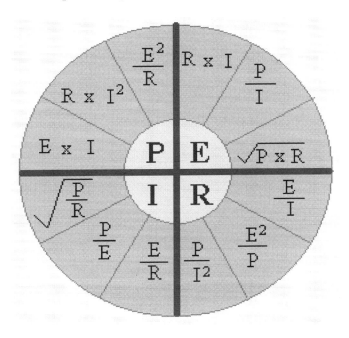

USEFUL FORMULAS

SERIES & PARALLEL RESISTANCE

If there is more than one resistor in series in a circuit the resistance is additive.

Inductance is also additive if in series in a circuit. However, capacitance is additive if the capacitors is in parallel

R1+R2+R3…

L1+L2+L3…

(if in parallel, capacitance follows the same formula of C1+C2+C3…)

In Parallel Resistance and Inductance follow a formula of reciprocals of

$$\cfrac{1}{\cfrac{1}{R1+R2+R3\ldots}}$$

AND

$$\cfrac{1}{\cfrac{1}{L1+L2+L3\ldots}}$$

However, Capacitance in parallel follows the formula of
C1+C2+C3…

INDUCTIVE REACTANCE

A.C. theory can be a little more difficult to grasp compared to D.C. theory. In an A.C. circuit current and voltage can be out of phase. That is to say they can occur at different times in a cycle. This means, since it takes both voltage and current working together to produce results, that the real and applicable wattage will be quite different than what may be measured. In an A.C. circuit an inductor or coil will resist a change in current. Because of this it pulls the current back by 90 degrees. In a purely inductive circuit E will lead I by 90 degrees. You can remember this by the name ELI.

The formula of $X_L = 2\pi F * L$. ($\pi = 3.1416$), f = frequency in MHz, L = inductance in henrys, gives the amount of impedance for an inductor in an A.C. circuit.

CAPACITIVE REACTANCE

In an A.C. circuit a capacitor will resist a change in voltage. This means it will cause current to be 90 degrees ahead of the voltage in a purely capacitive circuit. You can remember this with the word "ICE". Putting the memory tools of ELI and ICE together you have Eli the Ice man as a way to remember phasing.

The formula of capacitive reactance is $Xc = 1/([2\pi * f * c])$. So take two π (3.1416*2 or 6.2832) times frequency in MHz times capacitance in farads and then find the reciprocal of that and you have capacitive reactance. A capacitor will charge and discharge in units called Time Constants. It will charge to 63.2% in input voltage in one time constant and will discharge to 36.2% in one time constant.

Various components allow us to manipulate electricity.

Resistors impede the flow of electrons, acting to dissipate voltage. This resistance can be measured as heat.

Coils (also called inductors) store energy in a magnetic field. As the field changes the coil resists the change of current. This resistance or reactance occurs as the magnetic field of the coil contracts or expands, cutting across its own coil and thus acting like a small generator, which temporarily produces current counteracting the change. When a coil is in a circuit Voltage with lead Current by 90 degrees. The mnemonic is "ELI".

Capacitors store energy in an electrostatic field, building up electrons on one plate while a deficit forms on the other. As voltage changes the plate gives up electrons, thus capacitors resist a change in voltage. When a capacitor is in a circuit Current will lead Voltage by 90 degrees. Of course, when both a coil and capacitor are in a circuit the phase of current and voltage must be calculated. The mnemonic is "ICE".

Because a capacitor resists changes in voltage and coils resist changes in current they have little effect in a D.C. circuit because there is no change, but in an A.C. circuit the resistance they perform is called Reactance.

Inductors, or coils, drop voltage in proportion to the rate of current change. They will drop more voltage for faster-changing currents, and less voltage for slower-changing currents. What this means is that reactance in ohms for any inductor is directly proportional to the frequency of the alternating current. The exact formula for determining reactance is as follows:

$$X_L = 2\pi f L$$

As we discussed before, when an A.C. circuit has inductance and no capacitance in it Voltage will lead Current by 90 degrees. ELI

Capacitors resist a change in voltage. As the rate of voltage changes they will pass more current for faster-changing voltages (as they charge and discharge to the same voltage peaks in less time), and less current for slower-changing voltages. What this means is that reactance in ohms for any capacitor is *inversely* proportional to the frequency of the alternating current.

$$X_C = \frac{1}{2\pi f C}$$

As we discussed before, when a circuit has only capacitance and no inductance Current will lead Voltage by 90 degrees. ICE

(ELI the I

When both an inductor and capacitor is in a circuit the phase of voltage and current are effected by both and if there is more inductance (Henrys) than capacitance (Farads) in a circuit voltage will lead current. If there is more capacitance then current will lead voltage.

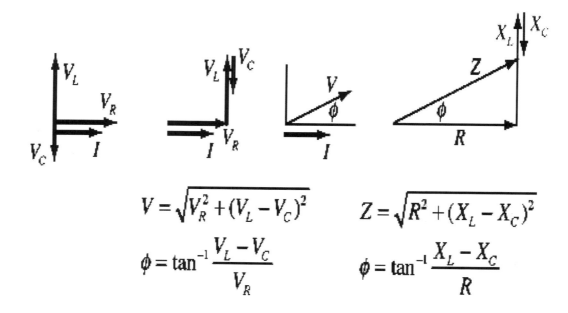

$$V = \sqrt{V_R^2 + (V_L - V_C)^2} \qquad Z = \sqrt{R^2 + (X_L - X_C)^2}$$

$$\phi = \tan^{-1}\frac{V_L - V_C}{V_R} \qquad \phi = \tan^{-1}\frac{X_L - X_C}{R}$$

As frequency increases a capacitor's reactance decreases and an inductor's reactance increases.

Another way to calculate phase is to use something called the "j-operator" This is an imaginary number described as the square root of negative 1 (-1). I know – this number cannot exist because no negative number can have a square root. Why? Because if you square a negative number you always get a positive number. However, it makes the formula work and is used in it as an imaginary number.

The j-operator has a value exactly equal to $\sqrt{-1}$, so successive multiplication of " j ", (j x j) will result in j having the following values of, -1, -j and +1. As the j-operator is commonly used to indicate the anticlockwise rotation of a vector, each successive multiplication or power of " j ", j^2, j^3 etc, will force the vector to rotate through an angle of 90° anticlockwise as shown below. Likewise, if the multiplication of the vector results in a -j operator then the phase shift will be -90°, i.e. a clockwise rotation.

Vector Rotation of the j-operator

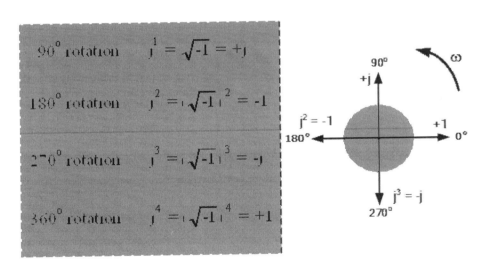

So multiplying the imaginary number j to the third power the vector will by 180° counterclockwise, multiplying by j^3 rotates it 270° and by j^4 rotates it 360° or back to its original position. Multiplication by j^{10} or by j^{30} will cause the vector to rotate counterclockwise by the appropriate amount. In each rotation, the magnitude of the vector always remains the same. The other way of visualizing these vectors is called the Cartesian or Rectangular method, which was exemplified with the tangent angles prior to this example.

When the capacitance and inductance in a circuit are tuned to the same frequency and that frequency is achieved the circuit is said to be resonant.

Resonance occurs when a system is able to store and easily transfer energy between two or more different storage modes such as a capacitor and inductor. Think of it as a child in a swing who kicks at the right time on each cycle thus going higher and higher. Gravity and the rocking of the body augment each other and the force grows. So it is with the components in a resonant circuit. Circuits of this type are used as tuning circuits and oscillators. Resonant circuits without feedback out of phase will run away and burn themselves up so part of the energy is fed back into the system 180 degrees out of phase to stabilize the oscillator.

Oscillator circuits are one of the basic building blocks of amateur radio equipment. Oscillator circuits are not only used to generate the signals we transmit. They are part of receivers, such as the superheterodyne receiver. In a superheterodyne receiver an oscillator generates a signal that is mixed with the signal we wish to receive. When signals are mixed the outcome is the two frequencies we start with, plus the sum and difference of those signals. So if a 10 kilohertz signal is mixed with a 7 kilohertz signal the output would have 10KHz, 7 KHz, 17 KHz, and 3KHz signals. We normally would use the lower frequency to work with to get the audio or information out of.

An oscillator is an amplifier with a tuned circuit feeding from its output back into the input. This tuned circuit might be an LC (coil and capacitor) circuit, also called a tank circuit, or it could be a crystal. The values of the components in the tuned circuit determine the output frequency of the oscillator. The reason we can think of it as an amplifier is that to be an oscillator it must not allow the energy to diminish so it must have a gain greater than 1 to keep itself going.

Transformers

A transformer is made of two or more coils. The magnetic flux produced by the Primary (P) coil cuts across the Secondary (S) coil, producing current flow The voltage induced on this, the *secondary* coil, depends on the number of turns in the wire of the coils in a relationship of the number of turns between the primary and secondary. To increase the output voltage from the input voltage you increase the amount of coils in the secondary compared to the primary. The opposite causes the voltage to decrease. Thus the formula is: Vp/Vs = Np/Ns

$$\frac{V_p}{V_s} = \frac{N_p}{N_s}$$

Where V is the voltage and N is the number of turns on the coil. Using this formula we can calculate the number of turns required to give a certain output voltage from a known input voltage.

If you feel more comfortable sticking with the symbol "E" for voltage, as I do, then substitute the symbols: Ep/Es = Np/Ns

But – remember – Power in must equal Power out, so as the voltage output increases the current must decrease.

Transformers use several types of cores to concentrate the flux and keep the energy between the coils. If a ferrite core is used then a multiplier of 1000 is used. If powdered iron is used then the multiplier is 100.

The formula for calculating the number of turns needed is N= 100 or 1000 X the square root of L/AL

N = number of turns

L = inductance in microhenrys

AL – inductance index in millihenrys per 100 turns.

Diodes

A diode is made when a positive and a negative semiconductors are joined. If a block of P-type semiconductor is placed in contact with a block of N-type semiconductor the result is of no value. We have two conductive blocks in contact with each other, showing no unique properties. The number of electrons is balanced by the number of protons in both blocks. Thus, neither block has any net charge.

The P-type material has positive majority charge carriers, holes, which are free to move about the crystal lattice. The N-type material has mobile negative majority carriers, electrons. Near the junction, the N-type material electrons diffuse across the junction, combining with holes in P-type material. The region of the P-type material near the junction takes on a net negative charge because of the electrons attracted. Since electrons departed the N-type region, it takes on a localized positive charge. The thin layer of the crystal lattice between these charges has been depleted of majority carriers, thus, is known as the *depletion region*. It becomes nonconductive intrinsic semiconductor material. In effect, we have nearly an insulator separating the conductive P and N doped regions. When a positive charge is applied to the P material and a negative charge is applied to the N material current will increase and as the voltage increases past 0.7 V (or .03 volts in germanium diodes), current increases considerably. Thus the diode conducts in one direction and blocks the flow of current in the other direction. By blocking electron flow in one direction AC can be turned into DC. By adding capacitors and inductors to smooth out the humps left where the alternating cycle was blocked we can build a power supply to supply DC. **The threshold voltage of a silicon diode is .7v and for the germanium is it .3v This may be on the test a time or two.**

In the diagram below we see a transformer used to lower the AC voltage and diodes used to block and direct AC so that the waveform is DC. This configuration of diodes is called a bridge and is used in most FULL WAVE RECTIFIERS. It allows 360 degrees of the wave to come through by directing both the positive and negative parts of the cycle in the same direction, forming all positive impulses, which are easier to filter. A capacitor and coil are used to smooth the waveform out from a series of humps, to pure DC.

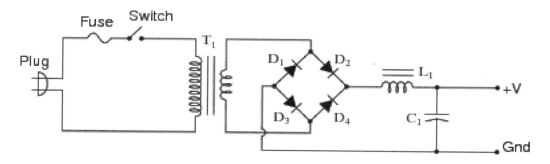

Here is another form of a full wave rectifier.

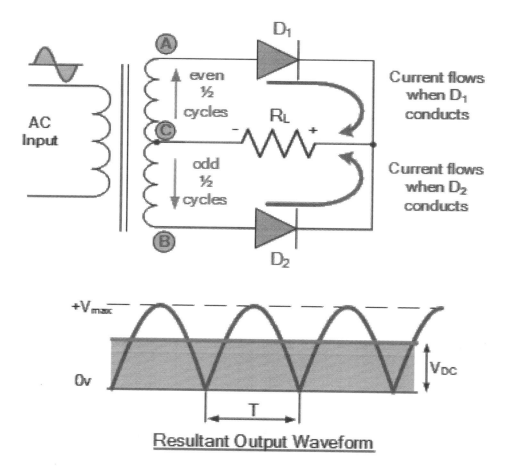

Resultant Output Waveform

As you can see, the full wave rectifiers produce impulses that are twice as close as the half wave rectifier below. The full wave rectifier converts 360 degrees of the wave where the half wave only uses 180 degrees.

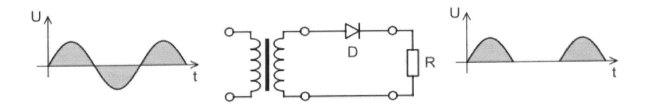

Transistor Theory

A transistor has two of the same type semiconductive materials, with a different type of semiconductive material sandwiched between them. Some of the material will have more electrons available and some will have fewer. If the two similar materials are P (having fewer available electrons) then the middle one must be N (having more available electrons). There are P-N-P transistors and N-P-N transistors.

Each transistor has 3 leads with a lead coming from each of its P or N elements. These elements are called the base, collector and emitter. The symbols of b, c and e are used respectively. The base is always the middle section.

The emitter will be designated by an arrow. If the transistor is a PNP the arrow will point in. The arrow points out for an NPN.

These are the symbols:

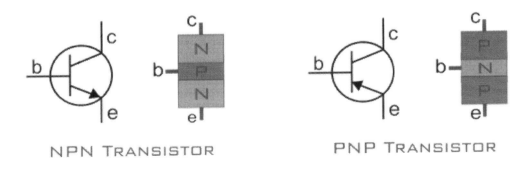

Transistor operation

In simple terms, when the N and P sections are placed together the charge is diffused until a steady and stable state is formed and no electron flow occurs. When a power source is connected between the base and the collector in reverse-bias, with the positive of the source connected to the collector and the negative to the base. The depletion zone of the P-N contact between the base and the collector will be widened as the electrons, which are negative, do not want to flow into the N material.

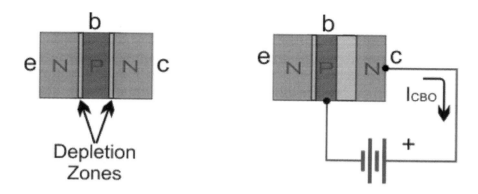

But when voltage is supplied between the emitter and the base in forward bias, with the positive of the source connected to the base and the negative connected to the emitter the depletion zone between the emitter and the base collapses, and electrons begin flow. The amount of voltage to collapse the zone depends on the material that the transistor is made of. For Germanium, the voltage is around 0.3 volts and for Silicon the voltage is around 0.7 volts. Some of the electrons that go through the e-b depletion zone, will fill the holes of missing electrons in the base. This is the base current.

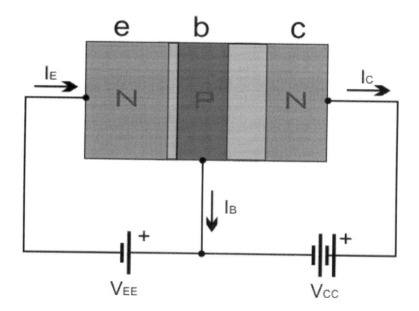

Electrons will flow through the base and will be directed to the collector. Electrons in the depletion zone between the base and the collector will be pushed by the voltage we connected to the transistor and will pass through the depletion zone. The electrons will then fill the holes in the collector, since that material lacks electrons. The electrons filling the holes will be replaced with holes coming from the base-collector positive voltage of the power supply (V_{CC}). The movement of these holes equals to a movement of electrons in the opposite direction, from the collector to the supply. This is called "hole flow" and is opposite of electron flow. Once past the obstacle presented by the second junction they are free to move through the N type material toward the second, even more positive, voltage source.

If we vary the current from the left electrode, the **emitter** E, to the center electrode, the **base** B, then a larger amount of current will flow to the right electrode, the **collector**. The variance of current will be proportional to the varying input to the base but the amplitude will be greater. The term "amplifier" is applicable since it rather allows us to *control* a large current with a small one. Thus, if we feed a small voltage in to a transistor and it causes a greater voltage to flow through the device we have amplified our change.

Although the mechanism is different the same control of flow and amplification is present in tubes with the Collector being equivalent to the Plate of a tube, the Emitter functioning like a Cathode, and the Base being like the Grid. In a tube a small amount of charge on the grid will affect a large change to the electron flow on the plate. Thus it amplifies.

The transistor that is most like a vacuum tube in its functionality is the FET – Field Effect Transistor.

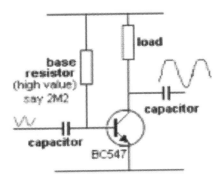

This is a "Common Emitter" amplifier. The capacitors block D.C. and only allow A.C. of certain frequencies to pass. Thus the capacitors isolate this stage from those before and after it.

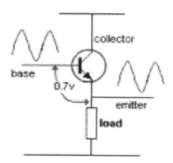

This is called an "Emitter Follower" amplifier. The output of the Emitter follows the input of the base. Not that the signals are in phase.

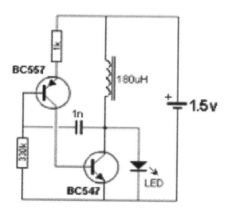

This is an Oscillator. The output of one stage is linked to the input of the other stage and then fed back into the input. The resistor in the feedback loop helps to keep it from allowing the signal to grow as it goes around and around. If it were allowed to get bigger each time the circuit would burn up. The frequency that feeds back in control by the resonant frequency of the coil. This sets the frequency of the oscillation.

Amplifiers are classified A, AB, B, C or D according to the phase-angle *(number of degrees of current flow during each 360-degree cycle)* over which collector-current flows.

Class A Amplifiers
Class A amplifiers operate over a relatively small portion of a tube's plate-current or a transistor's collector-current range and have continuous plate- or collector-current flow

throughout each RF cycle. Their efficiency in converting DC-source-power to RF-output-power is poor. DC source power that is not converted to radio frequency output power is dissipated as heat. However, in compensation, Class A amplifiers have greater input-to-output waveform linearity and less distortion than any other amplifier class. They are most commonly used in small-signal applications where linearity and a clean signal are more important than power efficiency. They can be used in high power signal applications where a clean signal is needed but the amp's disadvantages is poor power efficiency.

Class B Amplifiers
Class B amplifiers have their tube control-grids or transistor bases biased near plate- or collector-current cutoff, causing plate- or collector-current to flow only during approximately 180 degrees of each cycle. That causes output-power efficiency to be high but the output waveform can distort. Distortion is greatly reduced by using a high-Q resonant output "tank" circuit which will discharge at the frequency and reconstruct full cycles. This usually leaves distortion with part of the cycle smaller than the other. A more effective method commonly used to reduce Class B distortion is to make two amplifiers operate in "push-pull" such that one operates on one-half of the cycle and the other operate on the other half. The outputs are run together into a tank circuit to smooth switching transitions from the conduction of one amplifier to the other, and to correct other nonlinearities, creating a near-perfect, amplified waveform.

Class AB Amplifiers
The Class AB amplifier is a compromise between Class A and Class B operation. They are biased so plate- or collector-current flows less than 360 degrees, but more than 180 degrees, of each cycle. Any bias-point between those limits can be used, which provides a continuous selection-range extending from low-distortion, low-efficiency on one end to higher-distortion, higher-efficiency on the other.

Class AB amplifiers are widely used in SSB linear amplifier applications where low-distortion and high power-efficiency tend to both be very important.

Class C Amplifiers
Class C amplifiers are biased well beyond cutoff, so that plate- or collector-current flows less than 180 degrees of each RF cycle. That provides even higher power-efficiency than Class B operation, but with the penalty of even lower linearity and higher distortion. Making use of a high-Q resonant output tank circuit to restore complete RF sine-wave cycles is essential. High amplifying-nonlinearity makes them unsuitable to amplify AM, DSB, or SSB signals.

They also are commonly used in CW and frequency-shift-keyed radiotelegraph applications and in phase- and frequency-modulated transmitter applications where signal amplitudes remain constant.

Class D Amplifiers

Class D amplifiers are actually digital in nature. They work by generating a square wave of which the low-frequency portion of the spectrum is essentially the wanted output signal, and of which the high-frequency portion serves no purpose other than to make the wave-form binary so it can be amplified by switching the power devices.

A passive low-pass filter removes the unwanted high-frequency components, and smoothes the pulses out to recover the low-frequency signal. In other words, they chop[up the input signal into pieces within the cycle, digitize the pieces, amplifiy them and put them back together. What remains is the high frequency chopping artifacts, which are removed with a filter. The switching frequency is typically chosen to be ten or more times the highest frequency of interest in the input signal. This eases the requirements placed on the output filter.

This class of amplifier is a switching or PWM amplifier as mentioned above. This class of amplifier is the main focus of this application note. In this type of amplifier, the switches are either fully on or fully off, significantly reducing the power losses in the output devices. Efficiencies of 90-95% are possible. The audio signal is used to modulate a PWM carrier signal which drives the output devices, with the last stage being a low pass filter to remove the

high frequency PWM carrier frequency. From the above amplifier classifications, classes A, B and AB are all what is termed linear amplifiers.

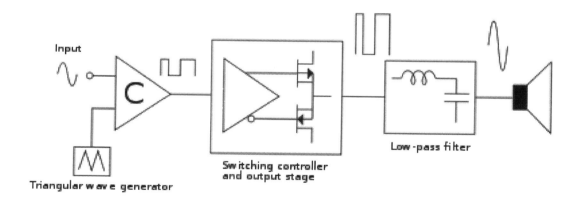

Excessive drive (input) to amplifiers can cause distortion of the output signal and damage to the amplifier. Flat-topping can occur where the amplifier clips off the top of the output wave form because it can provide no further amplification. This is a non-linear and distorted output.

An oscillator is an amplifier that has a feedback circuit, which keeps the oscillation from dying out. The frequency of the oscillation is determined by what frequencies are produced and allowed to feed back through the system. There are three types of oscillator circuits commonly used in Amateur Radio equipment. These are the **Colpitts, Hartley and Pierce** oscillator circuits.

Colpitts and Hartley oscillator circuits are commonly used in VFOs (variable frequency oscillators) because they can be easily manually tuned by use of a variable capacitor or coil.

In a Hartley oscillator, positive feedback is supplied **through a tapped coil**.

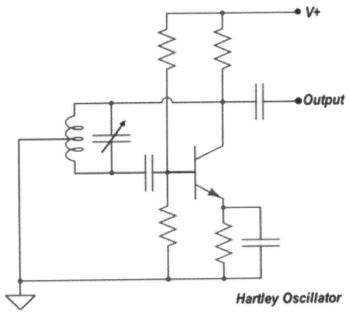

Hartley Oscillator

In a Colpitts oscillator, positive feedback is supplied **through a capacitor**.

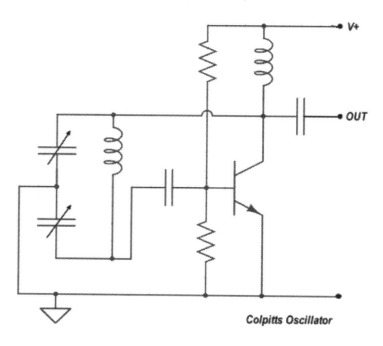

Colpitts Oscillator

In a Pierce oscillator, positive feedback is supplied **through a quartz crystal**.

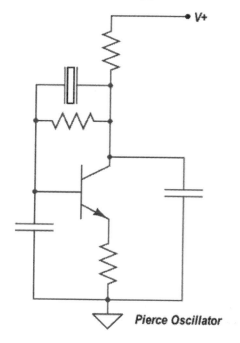

Pierce Oscillator

In addition to these basic oscillators, there are a couple of other oscillator types that you have to know about for the Extra Class test. A magnetron oscillator is **a UHF or microwave oscillator consisting of a diode vacuum tube with a specially shaped anode, surrounded by an external magnet**. A Gunn diode oscillator is **an oscillator**

based on the negative resistance properties of properly doped semiconductors. The diode turns on and off quickly enough to oscillate a very high frequency.

Oscillators produce the radio frequency signals, which are sent through power amplifiers and broadcast. These two simple stages make up a C.W. transmitter, where the signals are cut off and on to make dots and dashes.

Oscillators can be made to put out a sine wave, sawtooth wave, or a square wave. A sawtooth wave is used to make the electron beam of a TV picture tube or oscilloscope move across the screen to form a picture. The wave is fed into plates that form a charge that pushes the beam farther and farther across the screen, then cuts off and returns to the beginning and starts again. Another set of plates moves the beam up or down each time the cross scan begins. This forms interlacing scan lines from top to bottom making the picture we see on the screen. A square wave can be used as a pulse to time, clock, or turn circuits off and on.

When it comes to regular waves, like sine, square, and sawtooth, it is easy to calculate the RMS of those waves. We know that the RMS of a sine wave is .707 of peak. But when it comes to irregular waves, like the human voice, it is very difficult. The best way is to hook the output to a resistor of known value and measure the energy by measuring the heat coming off of the resistor. This tells us how much energy was absorbed by the resistor because it is re-radiated as heat.

The other thing to look at is the "duty cycle". RMS and Peak are values of the wave, but if an amplifier is working to amplify one wave a second or a million waves a second will make a tremendous difference in the ability of the amplifier to cool and last. The higher the frequency or greater the duty cycle the more constant the amplifier must work and the hotter it will get.

With the ability to make amplifiers, oscillators, and other stages radio technology began to develop and improve.

Designs for various stages improved and additions and evolutions followed. By making an amplifier with two inputs and one output a mixer is formed wherein the input frequencies are mixed. The result is the sum of the frequencies, the differences of the frequencies and the two original frequencies. By using an oscillator calibrated to change with a tuner a frequency can be produced that is always a certain amount of Hertz different than the received frequency. Mixing these two inputs together in a mixer stage would produce an output that is one stable frequency no mater what the frequency of the station being received. This mixing is called heterodyne.

Below is the block diagram showing the stages in a superheterodyne receiver.

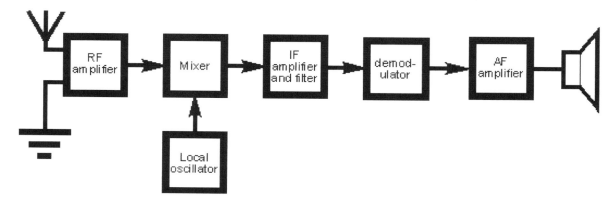

If the received station is transmitting in single sideband mode and the carrier is being suppressed, the receiver stage called a BFO or Beat Frequency Oscillator is added to mix with the ssb signal and reproduce the voice.

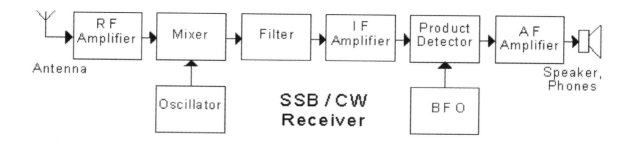

FM Receiver

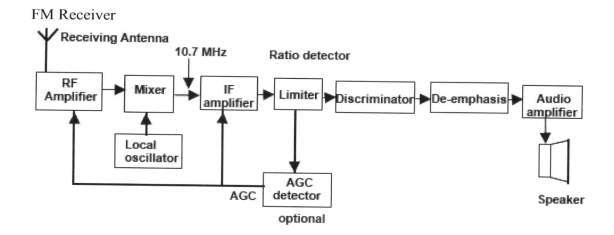

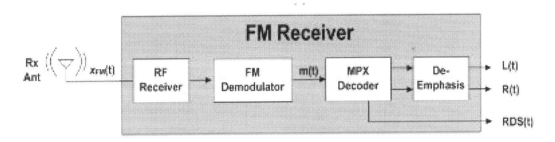

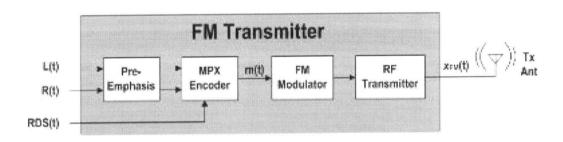

Since F.M. transmitters vary the frequency to convey intelligence this type of modulation is measured in frequency deviation or the modulation index. In the equation below, Δf is the frequency deviation, which represents the maximum frequency difference between the instantaneous frequency and the carrier frequency fm. In fact, the ratio of Δf to the carrier frequency is the modulation index. This index, β, is thus defined by the equation

$$\beta = \frac{\Delta f}{f_m}$$

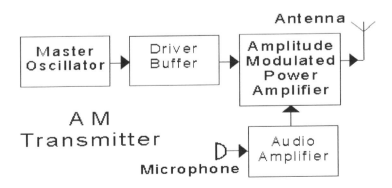

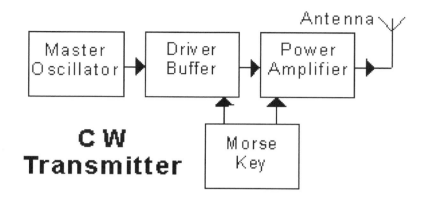

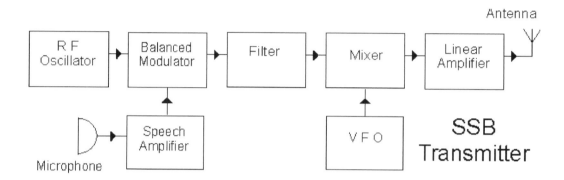

After the invention of the transistor a technique was developed to place transistors on a single substrate. These became known as "chips" or "Integrated Circuits (I.C.)". The TTL or transistor to transistor logic was reliable but demanded current and thus heated. Then the substrate was changed and CMOS was born.

Complementary metal–oxide–semiconductor (CMOS) is a technology for constructing integrated circuits. CMOS technology is used in microprocessors, microcontrollers, static RAM, and other digital logic circuits. CMOS technology is also used for several analog circuits such as image sensors (CMOS sensor), data converters, and highly integrated transceivers for many types of communication.

CMOS uses complementary and symmetrical pairs of p-type and n-type metal oxide semiconductor field effect transistors (MOSFETs) for logic functions.

Two important characteristics of CMOS devices are high noise immunity and low static power consumption. CMOS devices do not produce as much waste heat as other forms of logic, for example transistor-transistor logic (TTL) or NMOS logic, which normally have some standing current even when not changing state. CMOS also allows a high density of logic functions on a chip. It was primarily for this reason that CMOS became the most used technology to be implemented in Very Large Scale Integration (VLSI) chips.

There are other technologies in design and even those that combine TTY and CMOS when more power is needed in certain stages.

Frequency and Wavelength

There will be several questions on the Technicians test about frequency allocations and privileges. The exact spectrum allotted for certain licenses will have to be memorized, but those questions regarding frequency, wavelength, and antennas can be handled by a couple of formulas.

300/ frequency in MHz = wavelength in meters.

Remember that 1000 kilohertz = 1 megahertz.

Kilo means thousand and mega means million.

There is an easy to remember relationship between frequency and wavelength:

Prefixes tell how many zeros come before or after the unit. *Milli-* means thousandths,

so 1 milliAmpere (1 mA) is one–thousandth of an Ampere (.001 A).

Here's a chart with some common prefixes.

Prefix	Symbol	Power of 10	Multiplication Factor
giga	G	10^9	1,000,000,000
mega	M	10^6	1,000,000
kilo	K	10^3	1,000
milli	M	10^{-3}	0.001
micro	M	10^{-6}	0.000001
nano	N	10^{-9}	0.000000001
pico	p	10^{-12}	0.000000000001

An Expanded List of the Metric Prefixes

Prefix	Symbol	Multiplier Numerical	Exponential
peta	P	1,000,000,000,000,000	10^{15}
tera	T	1,000,000,000,000	10^{12}
giga	G	1,000,000,000	10^{9}
mega	M	1,000,000	10^{6}
kilo	k	1,000	10^{3}
hecto	h	100	10^{2}
deca	da	10	10^{1}
no prefix means:		1	10^{0}
deci	d	0.1	10^{-1}
centi	c	0.01	10^{-2}
milli	m	0.001	10^{-3}
micro	μ	0.000001	10^{-6}
nano	N	0.000000001	10^{-9}
pico	P	0.000000000001	10^{-12}

Look closely and you will see they progress by a factor of a thousand. One thousand times a factor will yield one of the next factors. The exceptions are 1, 10, and 100 or 1, .1, and .01.

Frequency allocations and bands are important to know. There are also general agreements you may be asked about. For example – Above 10 MHz single sideband communication is done on the Upper side band and below 10 MHz it is done on the lower side band. In satellite work you usually up link on the lower sideband and downlink on the upper sideband.

The term Telecommand refers to a one way communication which is meant to change or modify a system. Telemetry is a one way communication meant to deliver information or data.

There are 3 ways to control a station.

Local control, where the operator is present.

Remote control, where the operator has control of the station from a distant location.

Automatic control, where the station is run without an operator there or in immediate control via a setup or program.

Frequency Allocation according to class of License (ARRL Listing)

160 Meters

General, Advanced, Amateur Extra licensees:

1.800-2.000 MHz: CW, Phone, Image, RTTY/Data

80 Meters

Novice and Technician classes:

3.525-3.600 MHz: CW Only

General class:

3.525-3.600 MHz: CW, RTTY/Data
3.800-4.000 MHz: CW, Phone, Image

Advanced class:

3.525-3.600 MHz: CW, RTTY/Data
3.700-4.000 MHz: CW, Phone, Image

Amateur Extra class:

3.500-3.600 MHz: CW, RTTY/Data
3.600-4.000 MHz: CW, Phone, Image

60 Meters: Five Specific Channels

The FCC has granted hams secondary access on USB only to five discrete 2.8-kHz-wide channels. Amateurs cannot cause interference to and must accept interference from the Primary Government users. The NTIA says that hams planning to operate on 60 meters "must assure that their signal is transmitted on the channel center frequency."

This means that amateurs should set their carrier frequency 1.5 kHz *lower* than the channel center frequency.

General, Advanced and Amateur Extra classes:

Channel Center	Amateur Tuning Frequency
5332 kHz	5330.5 kHz
5348 kHz	5346.5 kHz
5358.5 kHz	5357.0 kHz
5373 kHz	5371.5 kHz
5405 kHz (common US/UK)	5403.5 kHz

Effective March 5, 2012, amateurs are permitted to use digital modes that comply with emission designator 60H0J2B, which includes PSK31 as well as any RTTY signal with a bandwidth of less than 60 Hz. They may also use modes that comply with emission designator 2K80J2D, which includes any digital mode with a bandwidth of 2.8 kHz or less whose technical characteristics have been documented publicly, per Part 97.309(4) of the FCC Rules. Such modes would include PACTOR I, II or III, 300-baud packet, MFSK16, MT63, Contestia, Olivia, DominoEX and others. with a maximum effective radiated power (ERP) of *100 W*. Radiated power must not exceed the equivalent of 100 W PEP transmitter output power into an antenna with a gain of 0 dBd.

40 Meters

Novice and Technician classes:

7.025-7.125 MHz : CW only

General class:

7.025-7.125 MHz : CW, RTTY/Data
7.175-7.300 MHz:: CW, Phone, Image

Advanced class:

7.025-7.125 MHz : CW, RTTY/Data
7.125-7.300 MHz:: CW, Phone, Image

Amateur Extra class:

7.000-7.125 MHz : CW, RTTY/Data
7.125-7.300 MHz:: CW, Phone, Image

Note: Phone and Image modes are permitted between 7.075 and 7.100 MHz for FCC licensed stations in ITU Regions 1 and 3 and by FCC licensed stations in ITU Region 2 West of 130 degrees West longitude or South of 20 degrees North latitude. See Sections 97.305(c) and 97.307(f)(11). Novice and Technician licensees outside ITU Region 2 may use CW only between 7.025 and 7.075 MHz and between 7.100 and 7.125 MHz. 7.200 to 7.300 MHz is not available outside ITU Region 2. See Section 97.301(e). These exemptions do not apply to stations in the continental US.

30 Meters

Maximum power, 200 watts PEP. Amateurs must avoid interference to the fixed service outside the US.

General, Advanced, Amateur Extra classes:

10.100-10.150 MHz: CW, RTTY/Data

20 Meters

General class:

14.025 -14.150 MHz CW, RTTY/Data
14.225 -14.350 MHz: CW, Phone, Image

Advanced class:

14.025 -14.150 MHz CW, RTTY/Data
14.175 -14.350 MHz: CW, Phone, Image

Amateur Extra class:

14.000 - 14.150 MHz CW, RTTY/Data
14.150 -14.350 MHz: CW, Phone, Image

17 Meters

General, Advanced, Amateur Extra classes:

18.068-18.110 MHz: CW, RTTY/Data
18.110-18.168 MHz: CW, Phone, Image

15 Meters

Novice and Technician classes:

21.025-21.200 MHz: CW Only

General class:

21.025-21.200 MHz: CW, RTTY/Data
21.275-21.450 MHz: CW, Phone, Image

Advanced class:

21.025-21.200 MHz: CW, RTTY/Data
21.225-21.450 MHz: CW, Phone, Image

Amateur Extra class:

21.000-21.200 MHz: CW, RTTY/Data
21.200-21.450 MHz: CW, Phone, Image

12 Meters

General, Advanced, Amateur Extra classes:

24.890-24.930 MHz: CW, RTTY/Data
24.930-24.990 MHz: CW, Phone, Image

10 Meters

Novice and Technician classes:

28.000-28.300 MHz: CW, RTTY/Data--Maximum power 200 watts PEP
28.300-28.500 MHz: CW, Phone--Maximum power 200 watts PEP

General, Advanced, Amateur Extra classes:

28.000-28.300 MHz: CW, RTTY/Data
28.300-29.700 MHz: CW, Phone, Image

6 Meters

All Amateurs except Novices:

50.0-50.1 MHz: CW Only
50.1-54.0 MHz: CW, Phone, Image, MCW, RTTY/Data

2 Meters

All Amateurs except Novices:
144.0-144.1 MHz: CW Only
144.1-148.0 MHz: CW, Phone, Image, MCW, RTTY/Data

1.25 Meters

The FCC has allocated 219-220 MHz to amateur use on a secondary basis. This allocation is *only* for fixed digital message forwarding systems operated by all licensees except Novices. Amateur operations must not cause interference to, and must accept interference from, primary services in this and adjacent bands. Amateur stations are limited to 50 W PEP output and 100 kHz bandwidth. Automated Maritime Telecommunications Systems (AMTS) stations are the primary occupants in this band. Amateur stations within 398 miles of an AMTS station must notify the station in writing at least 30 days prior to beginning operations. Amateur stations within 50 miles of an AMTS station must get permission in writing from the AMTS station before beginning operations. ARRL Headquarters maintains a database of AMTS stations. The FCC requires that amateur operators provide written notification including the station's geographic location to the ARRL for inclusion in a database at least 30 days before beginning operations. See Section 97.303(e) of the FCC Rules.

Novice (Novices are limited to 25 watts PEP output), Technician, General, Advanced, Amateur Extra classes:

222.00-225.00 MHz: CW, Phone, Image, MCW, RTTY/Data

70 Centimeters

All Amateurs except Novices:

420.0-450.0 MHz: CW, Phone, Image, MCW, RTTY/Data

33 Centimeters

All Amateurs except Novices:

902.0-928.0 MHz: CW, Phone, Image, MCW, RTTY/Data

23 Centimeters

Novice class:

1270-1295 MHz: CW, phone, Image, MCW, RTTY/Data (maximum power, 5 watts PEP)

All Amateurs except Novices:

1240-1300 MHz: CW, Phone, Image, MCW, RTTY/Data

Higher Frequencies:

All modes and licensees (except Novices) are authorized on the following bands [FCC Rules, Part 97.301(a)]:

2300-2310 MHz
2390-2450 MHz
3300-3500 MHz
5650-5925 MHz
10.0-10.5 GHz
24.0-24.25 GHz
47.0-47.2 GHz
76.0-81.0 GHz*
122.25 -123.00 GHz
134-141 GHz
241-250 GHz
All above 300 GHz

* Amateur operation at 76-77 GHz has been suspended till the FCC can determine that interference will not be caused to vehicle radar systems

Mode according to Frequency Allocation

160 Meters (1.8-2.0 MHz):

1.800 - 2.000	CW
1.800 - 1.810	Digital Modes
1.810	CW QRP
1.843-2.000	SSB, SSTV and other wideband modes
1.910	SSB QRP
1.995 - 2.000	Experimental
1.999 - 2.000	Beacons

80 Meters (3.5-4.0 MHz):

3.590	RTTY/Data DX
3.570-3.600	RTTY/Data
3.790-3.800	DX window
3.845	SSTV
3.885	AM calling frequency

40 Meters (7.0-7.3 MHz):

7.040	RTTY/Data DX
7.080-7.125	RTTY/Data
7.171	SSTV

| 7.290 | AM calling frequency |

30 Meters (10.1-10.15 MHz):

| 10.130-10.140 | RTTY |
| 10.140-10.150 | Packet |

20 Meters (14.0-14.35 MHz):

14.070-14.095	RTTY
14.095-14.0995	Packet
14.100	NCDXF Beacons
14.1005-14.112	Packet
14.230	SSTV
14.286	AM calling frequency

17 Meters (18.068-18.168 MHz):

| 18.100-18.105 | RTTY |
| 18.105-18.110 | Packet |

15 Meters (21.0-21.45 MHz):

| 21.070-21.110 | RTTY/Data |
| 21.340 | SSTV |

12 Meters (24.89-24.99 MHz):

24.920-24.925	RTTY
24.925-24.930	Packet

10 Meters (28-29.7 MHz):

28.000-28.070	CW
28.070-28.150	RTTY
28.150-28.190	CW
28.200-28.300	Beacons
28.300-29.300	Phone
28.680	SSTV
29.000-29.200	AM
29.300-29.510	Satellite Downlinks
29.520-29.590	Repeater Inputs
29.600	FM Simplex
29.610-29.700	Repeater Outputs

6 Meters (50-54 MHz):

50.0-50.1	CW, beacons
50.060-50.080	beacon subband
50.1-50.3	SSB, CW

50.10-50.125	DX window
50.125	SSB calling
50.3-50.6	All modes
50.6-50.8	Nonvoice communications
50.62	Digital (packet) calling
50.8-51.0	Radio remote control (20-kHz channels)
51.0-51.1	Pacific DX window
51.12-51.48	Repeater inputs (19 channels)
51.12-51.18	Digital repeater inputs
51.5-51.6	Simplex (six channels)
51.62-51.98	Repeater outputs (19 channels)
51.62-51.68	Digital repeater outputs
52.0-52.48	Repeater inputs (except as noted; 23 channels)
52.02, 52.04	FM simplex
52.2	TEST PAIR (input)
52.5-52.98	Repeater output (except as noted; 23 channels)
52.525	Primary FM simplex
52.54	Secondary FM simplex
52.7	TEST PAIR (output)

53.0-53.48	Repeater inputs (except as noted; 19 channels)
53.0	Remote base FM simplex
53.02	Simplex
53.1, 53.2, 53.3, 53.4	Radio remote control
53.5-53.98	Repeater outputs (except as noted; 19 channels)
53.5, 53.6, 53.7, 53.8	Radio remote control
53.52, 53.9	Simplex

2 Meters (144-148 MHz):

144.00-144.05	EME (CW)
144.05-144.10	General CW and weak signals
144.10-144.20	EME and weak-signal SSB
144.200	National calling frequency
144.200-144.275	General SSB operation
144.275-144.300	Propagation beacons
144.30-144.50	New OSCAR subband
144.50-144.60	Linear translator inputs
144.60-144.90	FM repeater inputs
144.90-145.10	Weak signal and FM simplex (145.01,03,05,07,09 are widely used for packet)

145.10-145.20	Linear translator outputs
145.20-145.50	FM repeater outputs
145.50-145.80	Miscellaneous and experimental modes
145.80-146.00	OSCAR subband
146.01-146.37	Repeater inputs
146.40-146.58	Simplex
146.52	National Simplex Calling Frequency
146.61-146.97	Repeater outputs
147.00-147.39	Repeater outputs
147.42-147.57	Simplex
147.60-147.99	Repeater inputs

Notes: The frequency 146.40 MHz is used in some areas as a repeater input. This band plan has been proposed by the ARRL VHF-UHF Advisory Committee.

1.25 Meters (222-225 MHz):

222.0-222.150	Weak-signal modes
222.0-222.025	EME
222.05-222.06	Propagation beacons
222.1	SSB & CW calling frequency

222.10-222.15	Weak-signal CW & SSB
222.15-222.25	Local coordinator's option; weak signal, ACSB, repeater inputs, control
222.25-223.38	FM repeater inputs only
223.40-223.52	FM simplex
223.52-223.64	Digital, packet
223.64-223.70	Links, control
223.71-223.85	Local coordinator's option; FM simplex, packet, repeater outputs
223.85-224.98	Repeater outputs only

Note: The 222 MHz band plan was adopted by the ARRL Board of Directors in July 1991.

70 Centimeters (420-450 MHz):

420.00-426.00	ATV repeater or simplex with 421.25 MHz video carrier control links and experimental
426.00-432.00	ATV simplex with 427.250-MHz video carrier frequency
432.00-432.07	EME (Earth-Moon-Earth)
432.07-	Weak-signal CW

432.10	
432.10	70-cm calling frequency
432.10-432.30	Mixed-mode and weak-signal work
432.30-432.40	Propagation beacons
432.40-433.00	Mixed-mode and weak-signal work
433.00-435.00	Auxiliary/repeater links
435.00-438.00	Satellite only (internationally)
438.00-444.00	ATV repeater input with 439.250-MHz video carrier frequency and repeater links
442.00-445.00	Repeater inputs and outputs (local option)
445.00-447.00	Shared by auxiliary and control links, repeaters and simplex (local option)
446.00	National simplex frequency
447.00-450.00	Repeater inputs and outputs (local option)

23 Centimeters (1240-1300 MHz):

Frequency Range	Suggested Emission Types

1240.00-1246.000	ATV
1246.000-1248.000	FM, digital
1248.000-1252.000	Digital
1252.000-1258.000	ATV
1258.000-1260.000	FM, digital
1240.000-1260.000	FM ATV
1260.000-1270.000	Various
1270.000-1276.000	FM, digital
1270.000-1274.000	FM, digital
1276.000-1282.000	ATV
1282.000-1288.000	FM, digital
1288.000-1294.000	Various
1290.000-1294.000	FM, digital
1294.000-1295.000	FM

	FM
1295.000-1297.000	
1295.000-1295.800	Various
1295.800-1296.080	CW, SSB, digital
1296.080-1296.200	CW, SSB
	CW, SSB
1296.200-1296.400	CW, digital
1296.400-1297.000	Various
1297.000-1300.000	Digital

Note: The need to avoid harmful interference to FAA radars may limit amateur use of certain frequencies in the vicinity of the radars.

13 Centimeters (2300-2310 and 2390-2450 MHz):

Frequency Range	Emission Bandwidth	Functional Use
2300.000-2303.000	0.05 - 1.0 MHz	Analog & Digital, including full duplex; paired with 2390 - 2393

2303.000-2303.750	< 50 kHz	Analog & Digital; paired with 2393-2393.750
2303.75-2304.000		SSB, CW, digital weak-signal
2304.000-2304.100	3 kHz or less	Weak Signal EME Band
2304.10-2304.300	3 kHz or less	SSB, CW, digital weak-signal (Note 1)
2304.300-2304.400	3 kHz or less	Beacons
2304.400-2304.750	6 kHz or less	SSB, CW, digital weak-signal & NBFM
2304.750-2305.000	< 50 kHz	Analog & Digital; paired with 2394.750 - 2395
2305.000-2310.000	0.05 - 1.0 MHz	Analog & Digital, paired with 2395 - 2400 (Note 2)
2310.000-2390.000		NON-AMATEUR
2390.000-2393.000	0.05 - 1.0 MHz	Analog & Digital, including full duplex; paired with 2300-2303
2393.000-2393.750	< 50 kHz	Analog & Digital; paired with 2303-2303.750
2393.750-2394.750		Experimental
2394.750-2395.000	< 50 kHz	Analog & Digital; paired with 2304.750 - 2305
2395.000-2400.000	0.05 - 1.0 MHz	Analog & Digital, including full duplex; paired with 2305-2310

2400.000-2410.000	6 kHz or less	**Amateur Satellite Communications**
2410.000-2450.000	22 MHz max.	**Broadband Modes (Notes 3, 4)**

Notes:
1: 2304.100 is the National Weak-Signal Calling Frequency
2: 2305 - 2310 is allocated on a primary basis to Wireless Communications Services (Part 27). Amateur operations in this segment, which are secondary, may not be possible in all areas.
3: Broadband segment may be used for any combination of high-speed data (e.g. 802.11 protocols), Amateur Television and other high-bandwidth activities. Division into channels and/or separation of uses within this segment may be done regionally based on needs and usage.
4: 2424.100 is the Japanese EME transmit frequency

For information, also called "intelligence", to be conveyed there must be a change in the signal through which information is encoded or transmitted. One can vary the amplitude of a signal. This is called AM. One can vary the frequency of a signal. This is called FM. One can vary the phasing of a signal. This is called PM. One can transmit a series of tones over AM or FM and let the change of tones encode the information. This frequency shift keying is called FSK. If a transmission is simply turned on and off with only the wave being started and stopped this is called continuous wave or CW. Each of these types of transmissions takes up "space" in the frequency range around the exact center frequency of the transmission. Some modes take up more room than others, much like a car would take up more room than a bike on the highway, or the human voice would take up more spectrum than a pure tone.

The FCC regulates the size of these bandwidths, and they can only be so wide. The FCC bandwidth allowance for RTTY, data, or multiplex on 6- and 2-meters is 20 kHz. Look at how *average* transmissions compare in the table below:

Mode	*Usual Bandwidth*
SSB Voice	2 - 3 kHz
FM Voice	10 - 20 kHz
Fast Scan	6 MHz

Television	

Part 97 places many different restrictions on how amateurs can use their stations and specifies technical standards that amateur radio station must meet.

Below is a chart from the ITU explaining this concept.

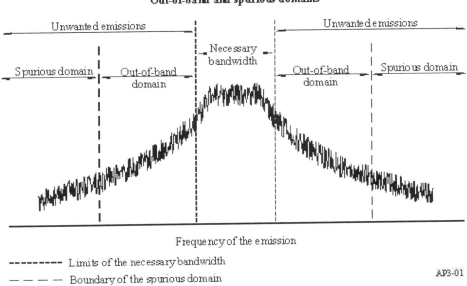

Here we see the upper side band, carrier center frequency, and the lower side band. If we suppress the carrier and one side band we can deliver much more power (usually about 3 times) into one side band.

YOU MUST KEEP ABOUT 3KHz away from the frequency boundaries of the bands otherwise your side bands will be illegal. Your center frequency will read 3 KHz higher on the lower end and 3 Khz lower on the higher end of the bands to stay legal.

Power is measured in watts, and references PEP (Peak Envelope Power) NOT RMS (Root Mean Square). WARNING! Most meters read RMS.

Peak power is measured from the center of the wave where it crosses zero to the peak.

Peak to peak is measured from the high peak to the low peak, so it is twice as much as peak power. RMS is measured from center to .707% down or up. RMS is .707 of peak. So, to find peak you multiply 1.414x RMS. To find peak to peak you multiply 2.828xRMS.

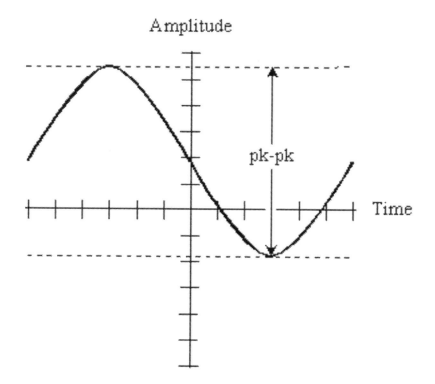

Peak x .707 = RMS
RMS X 1.414 = Peak
Peak to Peak = Peak X 2
Peak = .5 x Peak to Peak

1 dB = .79 decrease

1 dB increase = P x 1.26

1 dB decrease = P x .79

1 dB = .79 decrease

1 dB increase = P x 1.26

1 dB decrease = P x .79

Table of Electrical Symbols

Symbol	Component name
—/—	SPST Toggle Switch
—◆—	Solder Bridge
⏚	Earth Ground
⏛	Chassis Ground
⏣	Digital / Common Ground
Resistor Symbols	
—/\/\/—	Resistor (IEEE)
—▭—	Resistor (IEC)

〰️	Potentiometer (IEEE)
▭	Potentiometer (IEC)
〰️	Variable Resistor / Rheostat (IEEE)
⊙	Photo-resistor / Light dependent resistor (LDR)
⊣⊢	Capacitor
⊣⊢	Capacitor
+⊣⊢	Polarized Capacitor
+⊣⊢	Polarized Capacitor
⊣⊢	Variable Capacitor

Inductor / Coil Symbols	
⌐⁀⁀⁀⌐	Inductor
⌐⁀⁀⁀⌐ (with bar)	Iron Core Inductor
⌐⁀⁀⁀⌐ (with arrow)	Variable Inductor
	Coil / solenoid that generates magnetic field
⊖(- +)⊖	Voltage Source
⊖(→)⊖	Current Source
⊖(∿)⊖	AC Voltage Source
⊖(G)⊖	Generator
⊣├	Battery Cell
Battery	AC voltage source

○─⊕─○	Bulb	
○─⊖─○	Bulb	
Diode / LED Symbols		
○─▷○─○	NOT Gate (Inverter)	Outputs 1 when input is 0
═D─○	AND Gate	Outputs 1 when both inputs are 1.
═D○─	NAND Gate	Outputs 0 when both inputs are 1. (NOT + AND)
═D─○	OR Gate	Outputs 1 when any input is 1.
═D○─	NOR Gate	Outputs 0 when any input is 1. (NOT + OR)
═D─○	XOR Gate	Outputs 1 when inputs are different. (Exclusive OR)
[D Q]	D Flip-Flop	Stores one bit of data

Symbol	Name	Description
	Multiplexer / Mux 2 to 1	Connects the output to selected input line.
	Multiplexer / Mux 4 to 1	
	Demultiplexer / Demux 1 to 4	Connects selected output to the input line.

Symbol	Name	Description
	Diode	Diode allows current flow in one direction only (left to right).
	Zener Diode	Allows current flow in one direction, but also can flow in the reverse direction when above breakdown voltage
	Schottky Diode	Schottky diode is a diode with low voltage drop
	Varactor / Varicap Diode	Variable capacitance diode
	Tunnel Diode	
	Light Emitting Diode (LED)	LED emits light when current flows through

	Photodiode	Photodiode allows current flow when exposed to light
Transistor Symbols		
	NPN Bipolar Transistor	Allows current flow when high potential at base (middle)
	PNP Bipolar Transistor	Allows current flow when low potential at base (middle)
	JFET-N Transistor	N-channel field effect transistor
	JFET-P Transistor	P-channel field effect transistor
	NMOS Transistor	N-channel MOSFET transistor
	PMOS Transistor	P-channel MOSFET transistor

Logic gates are circuits that are designed to produce predetermined outputs with only specific inputs. An AND gate, for example, will only give a high output (3 – 5 volts) when both inputs are high at the same time. Otherwise the output remains low (zero volts).

These are the **symbols** for an AND gate, NOR gate and INVERTING or NOT gate. They have distinctive shapes making them easy to recognize.

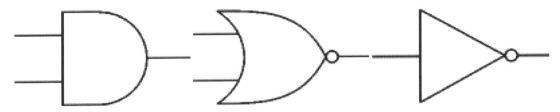

Gates have two or more inputs, except a NOT gate which has only one input. All gates have only one output. Usually the letters A, B, C and so on are used to label inputs, and Q is used to label the output. On this page the inputs are shown on the

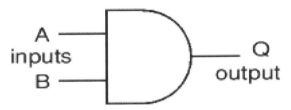

left and the output on the right. These are symbols for circuits that have specific binary outputs based on combinations of inputs. These predetermined results are called the Truth Table. For example, the AND gate to the right needs a high input on A and B for Q to be high. Any other combination will leave Q low. A low is zero volts and a high is between 3 and 5 volts.

The inverting circle (o)
Some gate symbols have a circle on their output which means that their function includes **inverting** of the output. It is equivalent to feeding the output through a NOT gate. For example the NAND (Not AND) gate symbol shown on the right is the same as an AND gate symbol but with the addition of an inverting circle on the output.

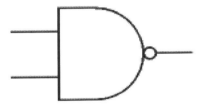

NOT gate (inverter)
The output Q is true when the input A is NOT true, the output is the inverse of the input: **Q = NOT A**
A NOT gate can only have one input. A NOT gate is also called an inverter.

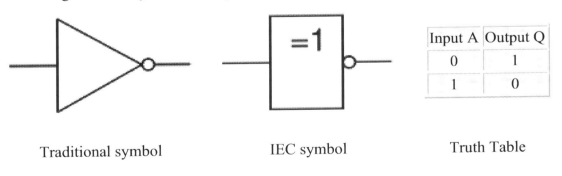

Traditional symbol   IEC symbol   Truth Table

AND gate
The output Q is true if input A and input B are both true: **Q = A and B**
An AND gate can have two or more inputs, its output is true if all inputs are true.

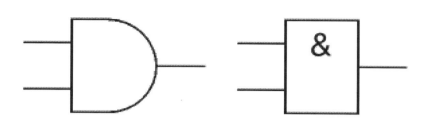

| Traditional symbol | IEC symbol | Truth Table |

Input A	Input B	Output Q
0	0	0
0	1	0
1	0	0
1	1	1

NAND gate (NAND = Not AND)
This is an AND gate with the output inverted, as shown by the 'o' on the output.
The output is true if input A AND input B are NOT both true: **Q = NOT (A AND B)**
A NAND gate can have two or more inputs, its output is true if NOT all inputs are true.

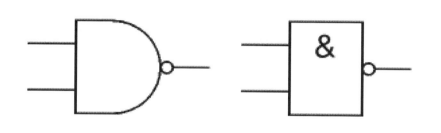

| Traditional symbol | IEC symbol | Truth Table |

Input A	Input B	Output Q
0	0	1
0	1	1
1	0	1
1	1	0

OR gate
The output Q is true if input A OR input B is true (or both of them are true): **Q = A OR B**
An OR gate can have two or more inputs, its output is true if at least one input is true.

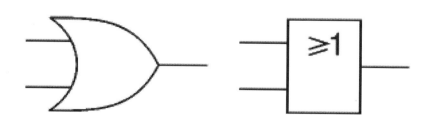

| Traditional symbol | IEC symbol | Truth Table |

Input A	Input B	Output Q
0	0	0
0	1	1
1	0	1
1	1	1

NOR gate (NOR = Not OR)
This is an OR gate with the output inverted, as shown by the 'o' on the output.
The output Q is true if NOT inputs A OR B are true: **Q = NOT (A OR B)**
A NOR gate can have two or more inputs, its output is true if no inputs are true.

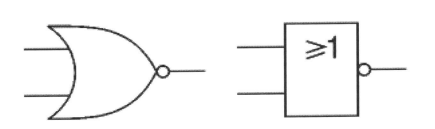

Input A	Input B	Output Q
0	0	1
0	1	0
1	0	0
1	1	0

Traditional symbol   IEC symbol   Truth Table

EX-OR (EXclusive-OR) gate
The output Q is true if either input A is true OR input B is true, **but not when both of them are true**: **Q = (A AND NOT B) OR (B AND NOT A)**
This is like an OR gate but excluding both inputs being true.
The output is true if inputs A and B are **DIFFERENT**.
EX-OR gates can only have 2 inputs.

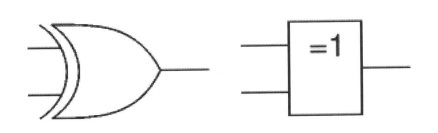

Input A	Input B	Output Q
0	0	0
0	1	1
1	0	1
1	1	0

Traditional symbol   IEC symbol   Truth Table

EX-NOR (EXclusive-NOR) gate
This is an EX-OR gate with the output inverted, as shown by the 'o' on the output.
The output Q is true if inputs A and B are the **SAME** (both true or both false): **Q = (A AND B) OR (NOT A AND NOT B)**
EX-NOR gates can only have 2 inputs.

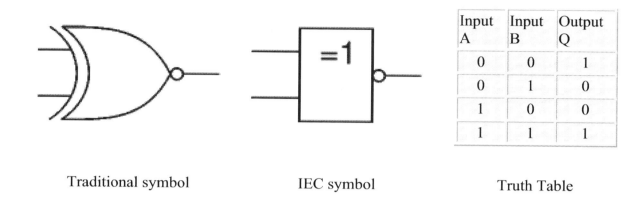

Input A	Input B	Output Q
0	0	1
0	1	0
1	0	0
1	1	1

Traditional symbol   IEC symbol   Truth Table

Circuits can be set up to be clocks and counters. There are circuits that can divide and multiply. By using flip-flop circuits a series of pulses can be seen on the input but a lesser number come out. If a flip-flop uses 2 pulses to change an output then by feeding 2 pulses in only a single pulse comes out.

Flip-flops can be divided into common types: the **SR** ("set-reset"), **D** ("data" or "delay," **T** ("toggle"), and **JK** types are the common ones. The behavior of a particular type can be described by what is termed the characteristic equation, which derives the "next" (i.e., after the next clock pulse) output, Qnext, in terms of the input signal(s) and/or the current output, Q.

JK flip-flop

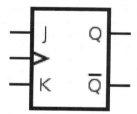

The JK flip-flop augments the behavior of the SR flip-flop (J=Set, K=Reset) by interpreting the S = R = 1 condition as a "flip" or toggle command. Specifically, the combination J = 1, K = 0 is a command to set the flip-flop; the combination J = 0, K = 1 is a command to reset the flip-flop; and the combination J = K = 1 is a command to toggle the flip-flop, i.e., change its output to the logical complement of its current value. Setting J = K = 0 does NOT result in a D flip-flop, but rather, will hold the current state. To synthesize a D flip-flop, simply set K equal to the complement of J. Similarly, to synthesize a T flip-flop, set K equal to J. The JK flip-flop is therefore a universal flip-flop, because it can be configured to work as an SR flip-flop, a D flip-flop, or a T flip-flop.

The characteristic equation of the JK flip-flop is: Qnext = J Qnot + Knot Q and the corresponding truth table is:

JK flip-flop operation							
Characteristic table				Excitation table			
J	K	Q_{next}	Comment	Q	Q_{next}	J	K	Comment
0	0	Q	hold state	0	0	0	X	No change
0	1	0	reset	0	1	1	X	Set
1	0	1	set	1	0	X	1	Reset
1	1	Q	toggle	1	1	X	0	No change

D flip-flop

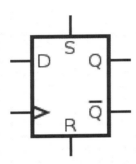

D flip-flop symbol

The D flip-flop is widely used. It is also known as a *data* or *delay* flip-flop.
The D flip-flop captures the value of the D-input at a definite portion of the clock cycle (such as the rising edge of the clock). That captured value becomes the Q output. At other times, the output Q does not change. The D flip-flop can be viewed as a memory cell, a zero-hold, or a delay line.

Clock	D	Q_{next}
Rising edge	0	0
Rising edge	1	1
Non-Rising	X	Q

('X' denotes a condition that does not matter, meaning the signal is irrelevant)
Most D-type flip-flops in ICs have the capability to be forced to the set or reset state (which ignores the D and clock inputs), much like an SR flip-flop. Usually, the illegal S = R = 1 condition is resolved in D-type flip-flops. By setting S = R = 0, the flip-flop can be used as described above.

Inputs				Outputs	
S	R	D	>	Q	Q'
0	1	X	X	0	1

| 1 | 0 | X | X | 1 | 0 |
| 1 | 1 | X | X | 1 | 1 |

T flip-flop

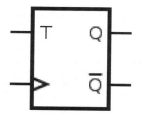

A circuit symbol for a T-type flip-flop

If the T input is high, the T flip-flop changes state ("toggles") whenever the clock input is strobed. If the T input is low, the flip-flop holds the previous value.

T flip-flop operation[28]							
Characteristic table				**Excitation table**			
T	Q	Q_{next}	Comment	Q	Q_{next}	T	Comment
0	0	0	hold state (no clk)	0	0	0	No change
0	1	1	hold state (no clk)	1	1	0	No change
1	0	1	toggle	0	1	1	Complement
1	1	0	toggle	1	0	1	Complement

When T is held high, the toggle flip-flop divides the clock frequency by two; that is, if clock frequency is 4 MHz, the output frequency obtained from the flip-flop will be 2 MHz. This "divide by" feature has application in various types of digital counters. A T flip-flop can also be built using a JK flip-flop (J & K pins are connected together and act as T) or D flip-flop (T input and $Q_{previous}$ is connected to the D input through an XOR gate). A T flip-flop can also be built using an edge-triggered D flip-flop with its D input fed from its own inverted output.

Astable multivibrator

An *astable circuit* is a form of oscillator. the word astable means unstable. An astable multivibrator consists of two active elements arranged in such way that the output of one is fed directly to the input of the other. Two identical resistance-capacitance networks determine the frequency at which oscillation will occur. The amplifying devices (tubes or transistors) are connected in a common-source or common-emitter configuration, as shown.

In the common-source or common-emitter circuit, the output of each transistor is 180 degrees out of phase with the input. An oscillating pulse might begin, for example, at the base of Q1 in the illustration. It is inverted at the collector of Q1, and goes to the base of

Q2. It is again inverted at the collector of Q2, and therefore returns to the base of Q1 in its original phase. This produces positive feedback, resulting in sustained oscillation.

The astable multivibrator is frequently used as an audio oscillator, but it is not often seen in RF applications because its output is extremely rich in harmonic products due to the shape of the waveform.

Monostable multivibrator

A monostable multivibrator is a circuit with only one stable condition. The circuit can be removed from this condition temporarily, but it always returns to that condition after a certain period of time. The monostable multivibrator is sometimes called a *one-shot multivibrator*.

Normally the output is high, at the level of supply voltage. When a positive triggering pulse is applied to the input, the output goes low(0V) for a length of time that depends on the values of the timing resistor R and the timing capacitor C. If R is given in Ohms and C is given in microseconds, time can be found by the equation:

$$T = 0.69RC$$

After the pulse duration time T has elapsed, the monostable multivibrator returns to the high state. Monostable multivibrators have been used as pulse generators, timing-wave generators and sweep generators for cathode-ray-tube devices.

Bistable multivibrator (flip-flop)

The bistable multivibrator is much more known as *flip-flop* because of its **two stable** switching states, hence flip, flop etc. It occurs when an input pulse of the right polarity and sufficient level is introduced. A two transistor the bistable multivibrator acts as a one bit memory.

Digital counter stage

There will be questions on the Extra Class tests regarding some types of gates.

Here is a logic table for all basic gates. The A and B inputs are represented in various states of high and low (power or no power). The resulting output is determined by what kind of gate is being used.

Summary for all 2-input gates							
Inputs		Output of each gate					
A	B	AND	NAND	OR	NOR	EX-OR	EX-NOR
0	0	0	1	0	1	0	1
0	1	0	1	1	0	1	0
1	0	0	1	1	0	1	0
1	1	1	0	1	0	0	1

Operational amplifiers, or "op-amps" have high input impedance and high gain.

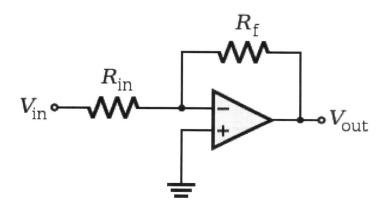

They are called ``operational" amplifiers, because they can be used to perform arithmetic operations (addition, subtraction, multiplication) with signals. In fact, op amps can also be used to integrate (calculate the areas under) and differentiate (calculate the slopes of) signals. The op-amp is basically a differential amplifier having a large voltage gain, very high input impedance and low output impedance. The op-amp has an "inverting" or (-) input and "noninverting" or (+) input and a single output.

The op-amp is connected using two resistors RA (R in) and RB (Rf) such that the input signal is applied in series with RA and the output is connected back to the inverting input through RB. The noninverting input is connected to the ground reference or the center tap of the dual polarity power supply. In operation, as the input signal moves positive, the output will move negative and visa versa. The amount of voltage change at the output relative to the input depends on the ratio of the two resistors RA and RB. As the input moves in one direction, the output will move in the opposite direction, so that the voltage at the inverting input remains constant or zero volts in this case. If RA is 1K and RB is 10K and the input is +1 volt then there will be 1 mA of current flowing through RA and the output will have to move to -10 volts to supply the same current through RB and keep the voltage at the inverting input at zero. The voltage gain in this case would be RB/RA or 10K/1K = 10. Note that since the voltage at the inverting input is always zero, the input signal will see a input impedance equal to RA, or 1K in this case.

The noninverting amplifier is connected so that the input signal goes directly to the noninverting input (+) and the input resistor RA is grounded. In this configuration, the input impedance as seen by the signal is much greater since the input will be following the applied signal and not held constant by the feedback current. As the signal moves in either direction, the output will follow in phase to maintain the inverting input at the same voltage as the input (+). **The voltage gain is always more than 1 and can be worked out from Vgain = (1+ RB/RA).**

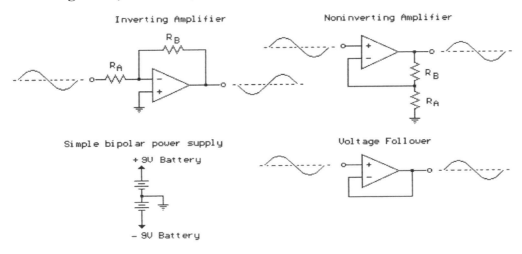

Modulation and Modes

So, now we have oscillators and amplifiers. With these devices we have the beginnings of a transmitter. We can produce radio frequency electro-magnetic waves with our oscillator and we can amplify them and send them to our antenna. But a simple wave can communicate no information. Information or intelligence can only be conveyed through change. Whether it is the changing lines of black upon white in the written word, the change of air pressure in waves with the spoken word, or the change of light intensity and frequency in sight, there must be a change to convey data.

When it comes to a transmitter the intelligence communicated can be conveyed in the radio waves being transmitted. There can be a simple, continuous wave, which is switched on and off. This binary approach is called Morse Code. Waves can be modulated by changing amplitude. Amplitude modulation is known as A.M. The frequency of the wave can be made to vary with speech. This frequency modulation is called F.M. The phase of the wave can be changed. This is called P.M. or phase modulation.

When an A.M. signal is transmitted the carrier can be suppressed and the power concentrated into the side bands. If only the center carrier is suppressed and both side bands are allowed this is called double side band. Usually only one side band is used and the carrier and one side band is suppressed. This is called single side band. It is the most efficient of the modes. Since the transmitter and R.F. amplifier only see energy when there is modulation and no carrier is present when there is no modulation the transmitter and amplifier can generate greater output without overheating.

Now that we see that the output must be varied to convey information the next step is to ask in what format will we vary the radio frequency waves? It is easy when it comes to voice. The voice will vary the output by using AM, FM, or PM. If we are using AM we can use only AM or we can use double side band or single side band, known as SSB. But there are great spaces of time unused when speaking, and there is no way to verify the content was correctly understood. The answer is to use digital modes. One may modulate in any of the above mentioned ways but what is sent becomes digital. Information is transmitted by shifting audio frequencies or tones. This is called Frequency Shift Keying. Each frequency shift or combination of shifts represents a digital state according to the protocol used.

An Overview of Digital HF Radio Operating Modes

TOR is an acronym for Teleprinting Over Radio. It is traditionally used to describe the three popular "error free" communication modes - AMTOR, PACTOR and G-TOR. The main method for error correction is from a technique called ARQ (Automatic Repeat Request) which is sent by the receiving station to verify any missed data. Since they share the same method of transmission (FSK), they can be economically provided together in one Terminal Node Controller (TNC) radio modem and easily operated with any modern radio transceiver. TOR methods that do not use the ARQ hand-shake can be easily operated with readily available software programs for personal computers. For the new and less complex digital modes, the TNC is replaced by an on-board sound card in the personal computer.

AMTOR is an FSK mode that is hardly used by radio amateurs in the 21st Century. While a robust mode, it only has 5 bits (as did its predecessor RTTY) and can not transfer extended ASCII or any binary data. With a set operating rate of 100 baud, it does not effectively compete with the speed and error correction of more modern ARQ modes like Pactor. The non-ARQ version of this mode is known as FEC, and known as SITOR-B by the Marine Information services.

PACTOR is an FSK mode and is a standard on modern Multi-Mode TNCs. It is designed with a combination of packet and Amtor Techniques. Although this mode is also fading in use, it is the most popular ARQ digital mode on amateur HF today and primarily used by amateurs for sending and receiving email over the radio. This mode is a major advancement over AMTOR, with its 200 baud operating rate, Huffman compression technique and true binary data transfer capability.

G-TOR (Golay -TOR) is an FSK mode that offers a fast transfer rate compared to Pactor. It incorporates a data inter-leaving system that assists in minimizing the effects of atmospheric noise and has the ability to fix garbled data. G-TOR tries to perform all transmissions at 300 baud but drops to 200 baud if difficulties are encountered and finally to 100 baud. (The protocol that brought back those good photos of Saturn and Jupiter from the Voyager space shots was devised by M.Golay and now adapted for ham radio use.) GTOR is a proprietary mode developed by Kantronics. Because it is only available with Kantronics multi-mode TNCs, it has never gained in popularity and is rarely used by radio amateurs.

PACTOR II is a robust and powerful PSK mode, which operates well under varying conditions. It uses strong logic, automatic frequency tracking; it is DSP based and as much as 8 times faster than Pactor. Both PACTOR and PACTOR-2 use the same protocol handshake, making the modes compatible. As with the original Pactor, it is rarely used by radio amateurs since the development of the new PC based sound card modes. Also, like GTOR, it is a proprietary mode owned by SCS and only available with their line of multi-mode TNC controllers.

CLOVER is a PSK mode which provides a full duplex simulation. It is well suited for HF operation (especially under good conditions), however, there are differences between CLOVER modems. The original modem was named CLOVER-I, the latest DSP based modem is named CLOVER-II. Clovers key characteristics are band-width efficiency with high error-corrected data rates. Clover adapts to conditions by constantly monitoring the received signal. Based on this monitoring, Clover determines the best modulation scheme to use.

RTTY or "Radio Teletype" is a FSK mode that has been in use longer than any other digital mode (except for morse code). RTTY is a very simple technique which uses a five-bit code to represent all the letters of the alphabet, the numbers, some punctuation and some control characters. At 45 baud (typically) each bit is 1/45.45 seconds long, or 22 ms and corresponds to a typing speed of 60 WPM. There is no error correction provided in RTTY; noise and interference can have a seriously detrimental effect. Despite its relative disadvantages, RTTY is still popular with many radio amateurs. This mode has now been implemented with commonly available PC sound card software.

PSK31 is the first new digital mode to find popularity on HF bands in many years. It combines the advantages of a simple variable length text code with a narrow bandwidth phase-shift keying (PSK) signal using DSP techniques. This mode is designed for "real time" keyboard operation and at a 31 baud rate is only fast enough to keep up with the typical amateur typist. PSK31 enjoys great popularity on the HF bands today and is presently the standard for live keyboard communications. Most of the ASCII characters are supported. A second version having four (quad) phase shifts (QPSK) is available that provides Forward Error Correction (FEC) at the cost of reduced Signal to Noise ratio. Since PSK31 was one of the first new digital sound card modes to be developed and introduced, there are numerous programs available that support this mode - most of the programs available as "freeware".

HF PACKET (300 baud) radio is a FSK mode that is an adaption of the very popular Packet radio used on VHF (1200 baud) FM amateur radio. Although the HF version of Packet Radio has a much reduced bandwidth due to the noise levels associated with HF operation, it maintains the same protocols and ability to "node" many stations on one frequency. Even with the reduced bandwidth (300 baud rate), this mode is unreliable for general HF ham communications and is mainly used to pass routine traffic and data between areas where VHF repeaters maybe lacking. HF and VHF Packet has recently enjoyed a resurgence in popularity since it is the protocol used by APRS - Automatic Position Reporting System mostly on 2 meter VHF and 30 meter HF.

HELLSCHREIBER is a method of sending and receiving text using facsimile technology. This mode has been around a long time. It was actually developed by Germany prior to World War II! The recent use of PC sound cards as DSP units has increased the interest in Hellschreiber and many programs now support this new...well I mean, old mode. The single-tone version (Feld-Hell) is the method of choice for HF operation. It is an on-off keyed system with 122.5 dots/second, or about a 35 WPM text rate, with a narrow bandwidth (about 75 Hz). Text characters are "painted" on the screen, as apposed to being decoded and printed. Thus, many different fonts can be used for this mode including some basic graphic characters. A new "designer" flavor of this mode called PSK HELL has some advantage for weak signal conditions. As with other "fuzzy modes" it has the advantage of using the "human processor" for error correction; making it the best overall mode for live HF keyboard communications. Feld-Hell also has the advantage of having a low duty cycle meaning your transmitter will run much cooler with this mode.

MT63 is a new DSP based mode for sending keyboard text over paths that experience fading and interference from other signals. It is accomplished by a complex scheme to encode text in a matrix of 64 tones over time and frequency. This overkill method provides a "cushion" of error correction at the receiving end while still providing a 100 WPM rate. The wide bandwidth (1Khz for the standard method) makes this mode less desirable on crowded ham bands such as 20 meters. A fast PC (166 Mhz or faster) is needed to use all functions of this mode. MT63 is not commonly used by amateurs because of its large bandwidth requirement and the difficulty in tuning in an MT63 transmission.

THROB is yet another new DSP sound card mode that attempts to use Fast Fourier Transform technology (as used by waterfall displays). THROB is actually based on tone pairs with several characters represented by single tones. It is defined as a "2 of 8 +1 tone" system, or more simply put, it is based on the decode of tone pairs from a palette of 9 tones. The THROB program is an attempt to push DSP into the area where other methods fail because of sensitivity or propagation difficulties and at the same time work at a reasonable speed. The text speed is slower than other modes but the author (G3PPT) has been improving his MFSK (Multiple Frequency Shift Keying) program. Check his web site for the latest developments.

MFSK16 is an advancement to the THROB mode and encodes 16 tones. The PC sound card for DSP uses Fast Fourier Transform technology to decode the ASCII characters, and Constant Phase Frequency Shift Keying to send the coded signal. Continuous Forward Error Correction (FEC) sends all data twice with an interleaving technique to reduce errors from impulse noise and static crashes. A new improved Varicode is used to increase the efficiency of sending extended ASCII characters, making it possible to transfer short data files between stations under fair to good conditions. The relatively wide bandwidth (316 Hz) for this mode allows faster baud rates (typing is about 42 WPM) and greater immunity to multi path phase shift. A second version called MFSK8 is available with a lower baud rate (8) but greater reliability for DXing when polar phase shift is a major problem. Both versions are available in a nice freeware Windows program created by IZ8BLY.

JT65 is intended for extremely weak but slowly-varying signals, such as those found on troposcatter or Earth-Moon-Earth (EME, or "moonbounce") paths. It can decode signals many decibels below the noise floor, and often allows amateurs to successfully exchange contact information without signals being audible to the human ear. Like the other digital modes, multiple-frequency shift keying is employed. However unlike the other digitalmodes, messages are transmitted as atomic units after being compressed and then encoded with a process known as forward error correction (or "FEC"). The FEC adds redundancy to the data, such that all of a message may be successfully recovered even if some bits are not received by the receiver. (The particular code used for JT65 is Reed-Solomon.) Because of this FEC process, messages are either decoded correctly or not decoded at all, with very high probability. After messages are encoded, they are transmitted using MFSK with 65 tones. Operators have also begun using the JT65 mode for contacts on the HF bands, often using QRP (very low transmit power usually less than 5 watts). While the mode was not originally intended for HF use, its popularity has resulted in several new programs being developed and enhancements to the original WSJT in order to facilitate HF operation.

Olivia was developed by Pawel Jalocha and is a ham radio digital mode designed to work in difficult (low s/n ratios plus multipath propagation) conditions on HF bands. The signal can be decoded even when it is 10-14 db below the noise floor (i.e. when the amplitude of the noise is slightly over 3 times that of the signal). It can also decode well under other noise, QSB, QRM, flutter caused by polar path propagation and even auroral conditions. Currently the only other digital modes that match or exceed Olivia in sensitivity are some of the WSJT program modes that include JT65A and JT65-HF which are certainly limited in usage and definitely not true conversation capable.

The standard Olivia formats (bandwidth/tones) are 125/4, 250/8, 500/16, 1000/32, and 2000/64. However the most commonly used formats in order of use are 500/16, 500/8, 1000/32, 250/8, and 1000/16. This can cause some confusion and problems with so many formats and so many other digital modes. After getting used to the sound and look of Olivia in the waterfall, though, it becomes easier to identify the format when you encounter it. About 90% of all current Olivia activity on the air is one of the 2 formats : 500/16 and 1000/32.

DominoEX is a digital mode using MFSK (Multi-Frequency Shift Keying), used to send data (for example, hand-typed text) by radio. MFSK sends data using many different tones, sent one at a time. Each tone element ('symbol') can carry several bits of data. Most other digital modes uses each tone to represent only one bit. Thus the symbol rate is much lower for the same data rate when MFSK is used. This is beneficial, since it leads to high sensitivity with good data rate and modest bandwidth. More importantly, low symbol rates are less effected by multi-path reception timing effects.

Therefore MFSK is ideal for HF operation since it has good noise rejection and good immunity to most propagation distortion effects which adversely affect reception of other modes. MFSK is already used on HF by modes such as MFSK16, ALE, THROB and Olivia, but DominoEX improves on the MFSK types of modes by employing an Incremental Frequency Keying strategy. DominoEX is also a reasonably narrow-band mode along the lines of MFSK16 or RTTY.

A narrow-band application of MFSK presents some challenges. The main problem is that radio transceivers with high stability and tuning accuracy are usually required, since very small frequency steps are used for example when compared with RTTY. MFSK is also prone to interference from data arriving from different ionospheric paths, and like many modes, it is prone to interference from fixed carriers within the data passband. Forward Error Correction (FEC) can be deployed to reduce errors, but such modes can become slow and difficult to operate or the modes consume an excessive amount of bandwidth. With DominoEX, a different approach was taken, concentrating on perfecting the design for best Near Vertical Incidence Signal or NVIS reception without requiring FEC. All the inherent MFSK problems are also avoided or much reduced.

DominoEX uses a series of new techniques to counter the general limitations of MFSK. To avoid tuning problems, IFK (Incremental Frequency Keying) is used, where the data is represented not by the frequency of each tone, but by the frequency difference between one tone and the next, an equivalent idea to differential PSK. An additional technique, called Offset Incremental Keying (IFK+) is used to manage the tone sequence in order to counter inter-symbol interference caused by multi-path reception. This gives the mode a great improvement in robustness.

Like Olivia above, there are several variations of the DominoEX mode: DominoEX4, DominoEX5, DominoEX8, DominoEX11, DominoEX16 and finally DominoEX22. The higher the number the faster the speed of transmission so in difficult conditions it may be wise to use the slower speed, while good conditions might allow for faster speeds.

Contestia is a digital mode directly derived from Olivia but not quite as robust. It is more of a compromise between speed and performance. It was developed by Nick

Fedoseev, UT2UZ who is also one of the key developers of the MixW Mult-digital mode software application used by many hams. Contestia sounds almost identical to Olivia, can be configured in as many ways, but has essentially twice the speed.

Contestia has 40 formats just like Olivia. The formats vary in bandwidth (125,250,500,1000, and 2000hz) and number of tones used (2,4,8,16,32,64,128, or 256). The most commonly used formats right now seem to be 250/8, 500/16, and 1000/32.

So just how well does Contestia perform under very weak signal conditions. Surprisingly well as it handles QRM, QRN, and QSB very easily. It decodes below the noise level but experience has shown that Olivia still outperforms Contestia depending on which variation of the modes are used. However, Contestia is twice as fast as Olivia on a given variation of each respective mode. It is an excellent weak signal, conversational, QRP, and long distance digital mode. When using it for keyboard to keyboard conversation under fair to good conditions, it can be more preferable to many hams than Olivia because of the faster speed.

Contestia gets an increase in speed by using a smaller symbol block size (32) than Olivia (64) and by a using 6-bit decimal character set rather than 7-bit ASCII set that Olivia does. Because it has a reduced character set and does not print out in both upper and lower case. Some traffic nets might not want to use this mode because it does not support upper and lower case characters and extended characters found in many documents and messages. For normal digital chats that does not pose any problem, but also because of these limitations, Contestia has not seen much use and is more of a novelty mode.

To speed up communications special three letter codes were invented for use with Morse Code. If one is sending dots and dashes by hand it is best to keep things short and simple. The codes are called Q codes because all of them start with the letter Q. They were so helpful we still use them today even with voice and digital communications. Here is a list of Q codes.

Code	Question	Answer or Statement
QRA	What is the name (or call sign) of your station?	The name (or call sign) of my station is ...
QRG	Will you tell me my exact frequency (or that of ...)?	Your exact frequency (or that of ...) is ... kHz (or MHz).

QRH	Does my frequency vary?	Your frequency varies.
QRI	How is the tone of my transmission?	The tone of your transmission is (1. Good; 2. Variable; 3. Bad)
QRJ	How many voice contacts do you want to make?	I want to make ... voice contacts.
QRK	What is the readability of my signals (or those of ...)?	The readability of your signal (or those of ...) is ... (1 to 5).
QRL	Are you busy?	I am busy. (or I am busy with ...) Please do not interfere.
QRM	Do you have interference?	I have interference.
QRN	Are you troubled by static?	I am troubled by static.
QRO	Shall I increase power?	Increase power.
QRP	Shall I decrease power?	Decrease power.
QRQ	Shall I send faster?	Send faster (... wpm)
QRS	Shall I send more slowly?	Send more slowly (... wpm)
QRT	Shall I stop sending?	Stop sending.
QRU	Have you anything for me?	I have nothing for you.
QRV	Are you ready?	I am ready.
QRW	Shall I inform ... that you are calling him on ... kHz (or MHz)?	Please inform ... that I am calling him on ... kHz (or MHz).
QRX	When will you call me again?	I will call you again at ... (hours) on ... kHz (or MHz)
QRZ	Who is calling me?	You are being called by ... on ... kHz (or MHz)

QSA	What is the strength of my signals (or those of ...)?	The strength of your signal (or those of ...) is ... (1 to 5).
QSB	Are my signals fading?	Your signals are fading.
QSD	Is my keying defective?	Your keying is defective.
QSG	Shall I send ... telegrams (messages) at a time?	Send ... telegrams (messages) at a time.
QSK	Can you hear me between your signals?	I can hear you between my signals.
QSL	Can you acknowledge receipt?	I am acknowledging receipt.
QSM	Shall I repeat the last telegram (message) which I sent you, or some previous telegram (message)?	Repeat the last telegram (message) which you sent me (or telegram(s) (message(s)) numbers(s) ...).
QSN	Did you hear me (or ... (call sign)) on .. kHz (or MHz)?	I did hear you (or ... (call sign)) on ... kHz (or MHz).
QSO	Can you communicate with ... direct or by relay?	I can communicate with ... direct (or by relay through ...).
QSP	Will you relay a message to ...?	I will relay a message to ... .
QSR	Do you want me to repeat my call?	Please repeat your call; I did not hear you.
QSS	What working frequency will you use?	I will use the working frequency ... kHz (or MHz).
QST	-	Here is a broadcast message to all amateurs.
QSU	Shall I send or reply on this frequency (or on ... kHz (or MHz))?	Send or reply on this frequency (or on ... kHz (or MHz)).
QSW	Will you send on this frequency (or on ... kHz (or MHz))?	I am going to send on this frequency (or on ... kHz (or MHz)).

QSX	Will you listen to ... (call sign(s) on ... kHz (or MHz))?	I am listening to ... (call sign(s) on ... kHz (or MHz))
QSY	Shall I change to transmission on another frequency?	Change to transmission on another frequency (or on ... kHz (or MHz)).
QSZ	Shall I send each word or group more than once?	Send each word or group twice (or ... times).
QTA	Shall I cancel telegram (message) No. ... as if it had not been sent?	Cancel telegram (message) No. ... as if it had not been sent.
QTC	How many telegrams (messages) have you to send?	I have ... telegrams (messages) for you (or for ...).
QTH	What is your position in latitude and longitude (or according to any other indication)?	My position is ... latitude...longitude
QTR	What is the correct time?	The correct time is ... hours
QTU	At what times are you operating?	I am operating from ... to ... hours.
QTX	Will you keep your station open for further communication with me until further notice (or until ... hours)?	I will keep my station open for further communication with you until further notice (or until ... hours).
QUA	Have you news of ... (call sign)?	Here is news of ... (call sign).
QUC	What is the number (or other indication) of the last message you received from me (or from ... (call sign))?	The number (or other indication) of the last message I received from you (or from ... (call sign)) is ...
QUD	Have you received the urgency signal sent by ... (call sign of mobile station)?	I have received the urgency signal sent by ... (call sign of mobile station) at ... hours.
QUE	Can you speak in ... (language), - with interpreter if necessary; if so, on what frequencies?	I can speak in ... (language) on ... kHz (or MHz).

QUF	Have you received the distress signal sent by ... (call sign of mobile station)?	I have received the distress signal sent by ... (call sign of mobile station) at ... hours.

Fundamentals of Antennas and Feed Lines

A word needs to be said here about coaxial cables or co-ax (coax). Even though radio waves travel at 300 million meters per second in free space, they do not do so in wire. The velocity factor in coax is the percentage of the speed of light that the radio waves achieve. Most inexpensive cable has a velocity factor of about 66%. Better cable can go up to 80% or better. Of course, as you slow the propagation of the waves down the wave length within the cable will increase. When cutting cable in relation to the wavelength this must be factored in.

As the frequency increases less and less of the energy will penetrate the wire and more energy will stay on the surface of the wire. This is called **Skin Effect.** It will increase the resistance of the wire to the waves at higher frequencies.

We have looked at oscillators and amplifiers. We have discussed modes and methods of conveying intelligence with radio frequencies. Now we are ready to transmit our signal. We need an antenna.

We know that when voltage is present in a wire in becomes a magnet. When AC voltage is fed into a wire, alternating current and voltage in the wire turns into alternating magnetic and electrostatic fields. This is the origin of radio waves. These waves are transmitted. Just like in a generator when a coil of wire cuts a magnetic field generating current, so any wire being cut by the transmitted energy will produce current and receive the signal. These are antennas used for transmitting and receiving. Usually they are the same antenna.

Radio waves travel in free space at the speed of light or about 300,000 kilometers per second (186,000 miles per second). If a radio signal in free space has a frequency of one second, the length of the wave will be 300,000 kilometers, or 186,000 miles.

If we were to increase the frequency to 10 cycles per second (Hertz) then each complete cycle would have traveled a length of only 30,000 kilometers, or 18,600 miles.

The speed of light divided by the number of cycles per second (Hertz) will yield the length of the wave or the wavelength. Since the speed of light is 300,000,000 meters per second, a 100,000,000 (100 MegaHertz) wave is 3 meters (about 10 feet) in length.

One of the simplest antennas is a quarter wave antenna. It is easy to look at the graph and see the simplicity of the ¼ wave dipole. The antenna will have two wires or elements with a ¼ wavelength wire used as the element to catch the wave. Look at the graph and you will see that one quarter way through the cycle the wave is at its first peak on the y axis at t1. You could also make the most of the wave using a half wave dipole when the wave is at the peak of t1 and lowest point at +5 on the x axis. This would be double the energy captured. Given, this is a simplistic way of looking at antennas but is does prove a point that there are only some lengths that work best and are resonant. This resonance is the PRIMARY point of antenna design.

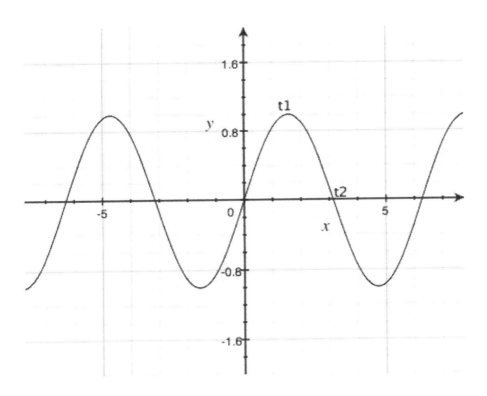

The dipole is an antenna made of metal or wire that is usually ¼ or ½ wavelength long. On the lower frequencies the dipole lengths become quite large. A dipole on the 80 meter band is about 130 feet in length for half a wavelength. (300 / 3.6 mhz) / 2 = 41.66 meters. 41.66 meters x 3.28 = 136.6 feet

Antenna Polarization

Since magnetic and electrostatic fields come off of an antenna in planes the antenna has polarization. The receiving antenna must be oriented in the same polarity as the transmitting antenna. Furthermore, the fields come off of the antenna broadside so that a wire stretched horizontally transmits and receives the signal on a horizontal plane and if the transmitting antenna is turned 90 degrees so the field of transmission is vertical so little of the field will cut across a horizontal receiving antenna that weak signals may not be heard at all. The diagram below shows a pattern typical of a half wave horizontal dipole.

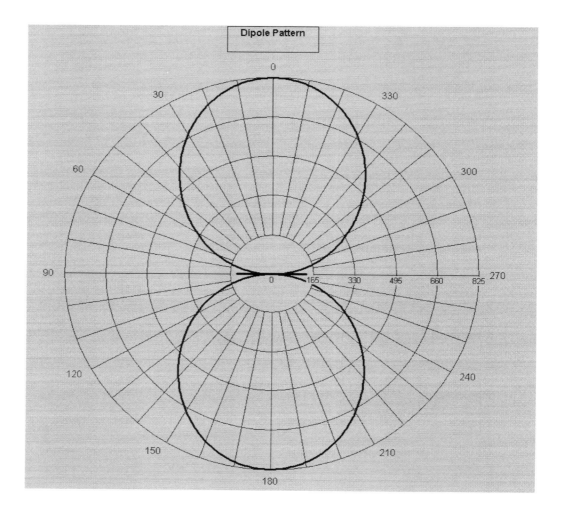

It is easy to see that if one were to be standing at the ends of the antenna there is little signal strength.

To design a dipole first decide what frequency you wish to work on, then calculate the length. The formula for calculating the antenna length in feet is 468 / Frequency (in MHz). If you wish to calculate the length in meters the formula is 300 / Frequency in

MHz. To convert to feet you can multiply meters by 3.28. For a dipole in the middle of the 10 meter band at 28.5 MHz the formula would be 468 / 28.5, or a length of 16.42 feet. Nothing is perfect and everything from ground height to masts will change the resonance slightly so cut the wire at least 10% longer. You can trim if needed.

The next thing to worry about regarding antennas is the impedance match between the antenna and the coax and radio. Coax is 50 – 52 ohms and radios are made to match at that impedance also. If there is a mismatch all the energy pushed out by the transmitter will not radiate. Some will be reflected back into the rig. Too much of this and it will damage the radio. The ratio between the energy output of the transmitter and that reflected back is called the Standing Wave Ratio or S.W.R.

If the radio is matched at 50 ohms and the feed point impedance is around 75 ohms the SWR will be 1.5 : 1.

An antenna may be resonant but may have impedance that does not match. This is because certain designs have impedance characteristics based on how the signal currents flow through the wire in various formations. A loop, for example, can be cut to frequency and be about 53 ohms. That is a good match. But, if that same loop is pulled at the sides so as to may it elliptical the impedance climbs. When the shape reaches that of a folded dipole it will reach about 300 ohms. This is good for a twin lead TV input, which expects 300 ohms but it is not good for a Ham Radio, which wants to see 50 Ohms. A 300 to 50 ohm match would make an SWR or 6:1 (300/50 = 6)

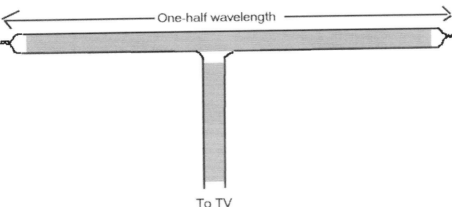

To match the antenna of one impedance with the line and radio of another impedance we will need a matching transformer or some device that will alter the impedance of the antenna.

ANTENNA MATCHING

An antenna may have an input impedance as low as 15 ohms or as high as 1000 ohms. However, most transmitters have an output impedance of 50 and most receivers these days are 50 or 75 ohms. Transmission lines are only available in a limited number of impedances of 50 or 75 ohms for coax and 300 ohms for flat line, also called twin lead.

All components, radio, line, and antenna, must be matched. There are a variety of matching techniques for antennas:

Lumped Match, also called Lumped Impedance or Lumped Component Match
Stub match
Delta match
Gamma Match
Quarter-wave transformer
LC network match
Transformer match

Lumped Match

One of the simplest of matches for dipoles is the series lumped component match, in which a coil or capacitor is placed in series or parallel with the antenna to bring the impedance of the antenna to the impedance of the coax. Lumped impedance matching can place coils or capacitors in series, parallel or both.

Stub Match

There is also a technique called line matching or stub matching. In stub matching a short piece of coax with a length of less than ¼ wavelength is placed between the antenna and feed line to the radio. If a stub of coax less than ¼ wavelength is used it can compensate for antenna impedances that are there are either capacitive or inductive in value. If one end is placed in line and the other end is shorted the stub will be inductive in value, while an open stub of similar length would be capacitive.

Hair Pin Match

The hair pin match is a type of stub match. The Beta or Hairpin Match is a simple and robust form of matching a lower impedance Yagi to the transmission line.

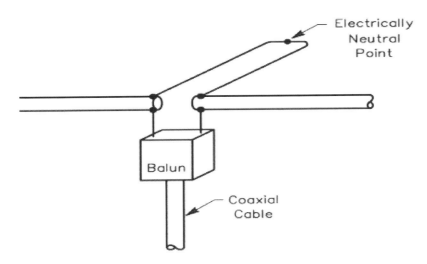

The diagram above adapted from the ARRL Antenna Handbook shows a Beta Match in the centre of a Yagi Driven Element.

The system operates by shortening the driven element so that the parallel equivalent impedance at the DE terminals is the desired load resistance in shunt with some capacitive reactance. The hairpin offers an inductive reactance that is connected in parallel with the driven element to cancel that capacitive reactance. The Beta or Hairpin match depends on shortening the driven element to change the feedpoint. So, this short circuit stub (the hairpin) needs to have a high Zo to keep its losses low.

Gamma Match

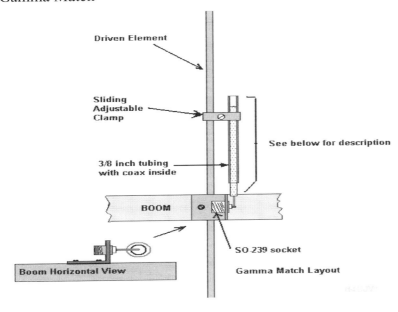

The Gamma Match is a capacitive matching system.

In the case above the center of the coax is inserted into the 3/8 inch aluminum pipe. The plastic center provides the dielectric. The tube and center conductor wire form the plates. The gamma match capacitor can only cancel reactance (impedance). The match cannot modify the actual resistance of the antenna. The Gamma match is the simplest form of matching, when it is useable.

Delta Match

The delta match uses a section of 2-wire transmission line with gradually increasing separation to match a 2-wire line to an antenna. Two advantages of the delta match is the simplicity of its construction and its ability to match a wide range of impedances.

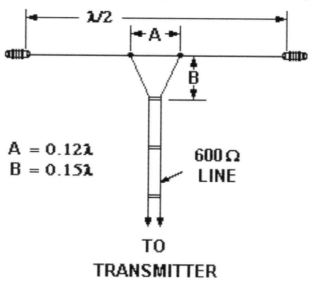

A variation of the delta match that is widely used for VHF/UHF Yagi antennas is the T-match:

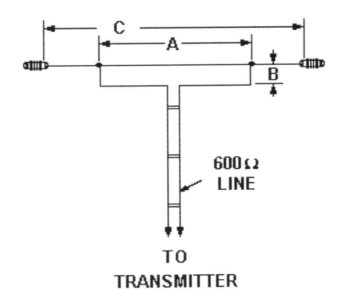

The T-match, unlike the delta match, does not radiate, but this advantage is offset by the need to adjust the overall length of the antenna slightly, as well as the T-match separation and length, to get a proper match. In general, lengths A, B and C must be determined by experiment.

QUARTER WAVE TRANSFORMER

The quarter wave transformer is a quarter wavelength section of transmission line whose characteristic impedance is selected to provide a match between the antenna and the main transmission line. When you calculate for the length of the coaxial cable you must remember that the velocity of radio waves is slower through wire than through the air. Usually the velocity is in the area of 65% - 75%.

The quarter wave transformer can theoretically be used to match any antenna impedance to any feed line impedance, although it is difficult to construct quarter wave line sections to match two low values of impedance. The impedance, Z_Q of the quarter wave line necessary to match an antenna of impedance Z_A to a feed line of impedance Z_0 is given by:

$$Z_Q = \sqrt{Z_0 Z_A}$$

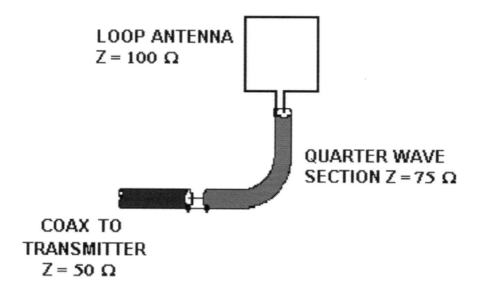

The advantages of the quarter wave transformer are that it is very easy to construct, and it can be used to transform a wide range of impedances. The disadvantages are that it is only useful over a narrow bandwidth and that a transmission line of proper characteristic impedance may not be available.

The narrow bandwidth of the match is the result of the requirement that the matching section be one quarter wavelength long. As we move the operating frequency away from the design frequency, the electrical length of the line changes and the matching section is no longer the proper length.

Transmission lines are commercially available in a limited number of characteristic impedances: 50, 75, 95, 135, 300, and 450 ohms. If a different impedance is required for the matching section, the matching section must be hand made. Depending on the impedance required, it may not be possible to construct a line with the proper impedance.

LC NETWORK MATCH

The LC network match consists of a network of capacitors and inductors that are used to transform the antenna impedance into the feed line impedance. There are three types of LC matching networks in wide use:

    L-network

    T-network

p-network

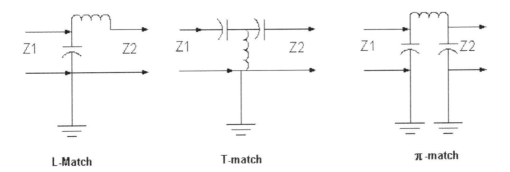

L-Match   T-match   π-match

The advantages of this type of matching are that any two values of impedance may be matched and there are formulas available that permit computation of all component values necessary to achieve a match. The major disadvantage is that the network will only match the impedances over a relatively narrow bandwidth.

We will examine only the equations for the L-Match because the mathematics for the other two networks is considerably more complicated. Consider the diagram below. It shows an L-network that is matching two purely resistive impedances, Rp and Rs. The two circuit elements, Xp and Xs may be either capacitors or inductors at the discretion of the circuit designer, but if one, say Xp is chosen to be inductive, the other should be capacitive.

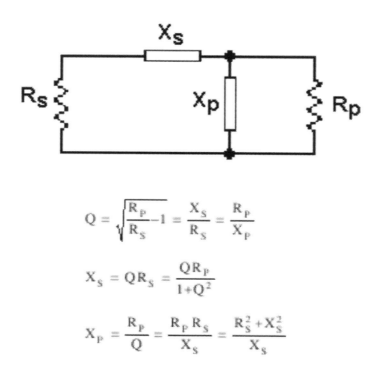

$$Q = \sqrt{\frac{R_P}{R_S} - 1} = \frac{X_S}{R_S} = \frac{R_P}{X_P}$$

$$X_S = QR_S = \frac{QR_P}{1+Q^2}$$

$$X_P = \frac{R_P}{Q} = \frac{R_P R_S}{X_S} = \frac{R_S^2 + X_S^2}{X_S}$$

Here is an example of L-network design:

We wish to match an antenna whose impedance is 250 ohms to a 50 ohm coaxial cable.

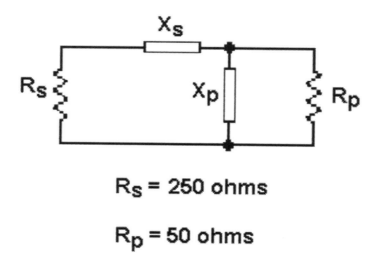

Rs = 250 ohms

Rp = 50 ohms

The first step is to compute the network Q:

$$Q = \sqrt{\frac{R_s}{R_p} - 1} = \sqrt{\frac{250}{50} - 1} = \sqrt{5-1} = \sqrt{4} = 2$$

Next we compute the series and parallel reactances, Xs and Xp:

$$X_s = QR_s = 2*500 = 1000$$

$$X_p = \frac{R_p}{Q} = \frac{50}{2} = 25$$

Now we are finished. Depending on which component is chosen to be a capacitor, there are two possible matching networks:

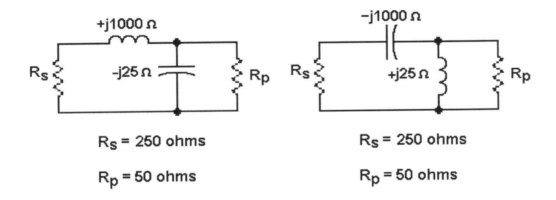

Both networks will make the required match. Normally, the network that yields the smallest component values is chosen. To see how this works, let us make the operating frequency for this network 2.4 MHz. We can then use the formulas for capacitive and inductive reactance to determine the component values.

For the network with the series inductance:

$$L_s = \frac{X_s}{2\pi f} = \frac{1000\Omega}{2\pi * 2.4MHz} = \frac{1000\Omega}{2\pi * 2.4MHz} = \frac{1000}{15.08} = 66.3\mu H$$

$$C_p = \frac{1}{2\pi f X_p} = \frac{1}{2\pi * 2.4MHz * 25\Omega} = \frac{1}{376992000} = 2654 pF$$

For the network with the parallel inductance:

$$L_p = \frac{X_p}{2\pi f} = \frac{25\Omega}{2\pi * 2.4MHz} = \frac{25\Omega}{2\pi * 2.4MHz} = \frac{25}{15.08} = 1.66\mu H$$

$$C_s = \frac{1}{2\pi f X_s} = \frac{1}{2\pi * 2.4MHz * 1000\Omega} = \frac{1}{15079680000} = 66.3 pF$$

In this case, the network with the parallel inductance has the smaller component values, so it would be the best choice in most circumstances.

It is possible to combine L, T or p networks together to make matching networks with almost any desired band pass response and input and output impedance. The analysis of these networks is quite complex and will not be covered in this course.

TRANSFORMER MATCHING

Matching using a transformer needs a specially designed RF transformer to match the antenna to the transmission line.. Its chief advantage is that it is a broadband matching device. Its chief disadvantage is that it does not work well with extremely large impedances (> 600 ohms)

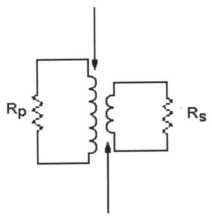

The RF transformer works very much like its low frequency counterpart. The relationship between the number of turns in each winding and the impedance ratio is given by:

$$\frac{Z_S}{Z_P} = \left(\frac{N_S}{N_P}\right)^2$$

Because they work over a wide frequency range, RF transformers are often used for impedance matching. The turns may be wound over a hollow core or may be wound onto a toroid made from powdered iron or a ferrite.

A Smith chart may help with phasing issues.

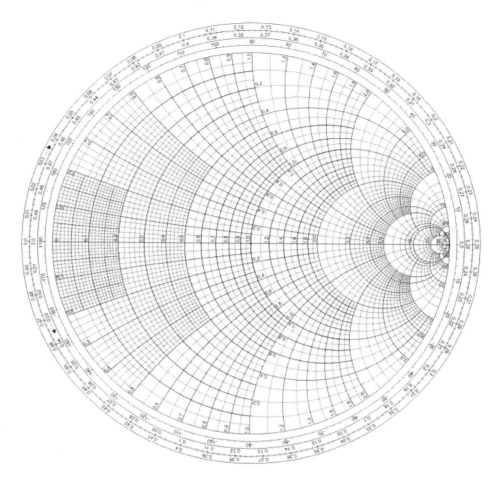

The Smith chart, invented by Phillip H. Smith (1905–1987), is a graphical aid or nomogram designed for electrical and electronics engineers specializing in radio frequency (RF) engineering to assist in solving problems with transmission lines and matching circuits. Use of the Smith chart utility has grown steadily over the years and it is still widely used today, not only as a problem solving aid, but as a graphical demonstrator of how many RF parameters behave at one or more frequencies, an alternative to using tabular information. The Smith chart can be used to represent many parameters including impedances, admittances, reflection coefficients, $S_{nn}\,$, scattering parameters, noise figure circles, constant gain contours and regions for unconditional stability, including mechanical vibrations analysis. The Smith chart is most frequently used at or within the unity radius region. However, the remainder is still mathematically relevant, being used, for example, in oscillator design and stability analysis.

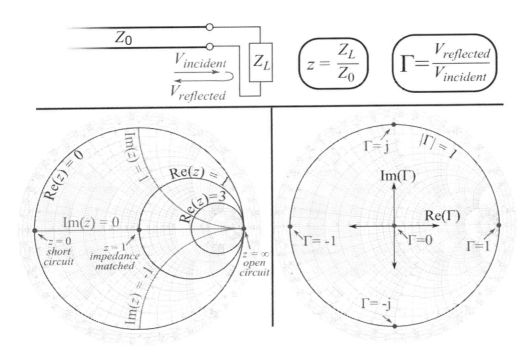

The SI unit of impedance is the ohm with the symbol of the upper case Greek letter Omega (Ω) and the SI unit for admittance is the siemens with the symbol of an upper case letter S. Normalized impedance and normalized admittance are dimensionless. Actual impedances and admittances must be normalized before using them on a Smith chart. Once the result is obtained it may be de-normalized to obtain the actual result.

The normalized impedance Smith chart

Using transmission line theory, if a transmission line is terminated in an impedance (Z_T,) which differs from its characteristic impedance (Z_0,), a standing wave will be formed on the line comprising the resultant of both the forward (V_F,) and the reflected (V_R,) waves. Using complex exponential notation:

Here is an illustration of the free space figure 8 pattern of a dipole.

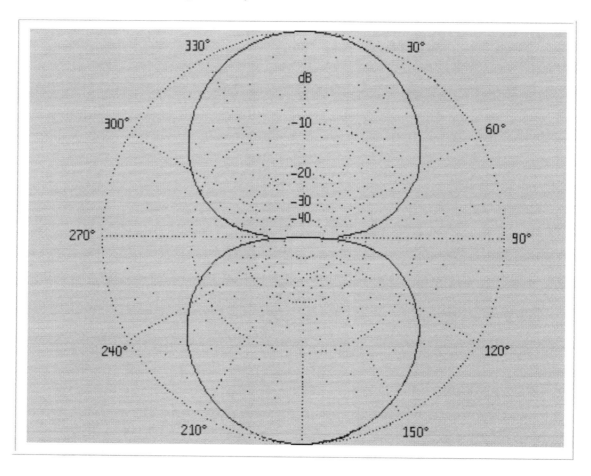

GENERAL DESCRIPTION OF A YAGI

The word "Yagi" is used to describe a type of antenna and is credited to very famous Japanese antenna experts by the names of Yagi and Uda! Most hams refer to this type of antenna as the "Yagi" rather than use both men's names.

They discovered that by adding "elements" of various lengths and spacings in front of and behind a dipole antenna that the performance and effectiveness of the dipole could be greatly increased and the pattern of the dipole rf energy could be "beamed" or focused in one direction, with the resulting "effect" of making it appear that the transmitter was running lots more power than it actually was, yielding much stronger signals both on receive and transmit!

The Yagi antenna's overall basic design consists of a "resonant" fed dipole and other parasitic elements. These parasitic elements are called the "reflector" and the "directors."
The reflector is behind the driven element and the directors are in from of the driven element.

You can have a 2 element Yagi using a dipole for the driven or radiating element and a reflector. There does not have to be a director. Such a simple 2 element Yagi was modified and developed into the "Hex Beam".

The hexagonal beam offers a number of features as follows:

- Gain and front/back comparable to a two element yagi.
- Six bands with low SWR
- Broad band characteristics
- Low weight and low wind load
- Construction from general hardware components
- Ease of adjustment

If you have been to other sites on construction of the hexagonal beam you might be a bit confused. You see, some sites tell you how to build the "original" hexagonal beam which is patterned after the design of the HEX-BEAM, a trade marked product of Traffie Technologies. The wires for this original design for a single band look from above, like an "M" over a "W".

For a more full understanding of the technical parameters of the G3TXQ broad band hexagonal beam, visit the web site of its inventor, Steve Hunt, G3TXQ.

If you feel you would rather not get into building your own G3TXQ broad band hexagonal beam, I can build one for you. See the Hexagonal Beam by K4KIO for sale details here.

This original design is a good antenna and owners of the HEX-BEAM are quite vocal about its performance as were builders of the homebrew version. I used to be one of the homebrew builders and was so enthusiastic that I published a set of guidelines like these to help others build one.

But, things have progressed a little and thanks to the exhaustive work of Steve Hunt, G3TXQ, a slightly different configuration of the hexagonal beam has been discovered. Viewed from above the wires for a single bander look like the sketch.

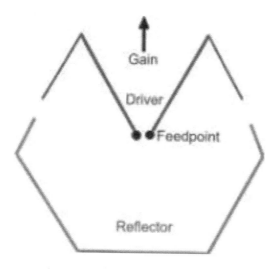

However, the basic concept of a Yagi is an antenna consisting of 3 or more elements. These include a driven element with a reflector behind it and one or more directors in front of the driven element.

From experimentation, they determined that the "effect" of their designs created much more "powerful" antennas compared to the standard dipole by just adding a few more elements to it. They also learned that by changing the space between the elements, and the element lengths, that they could "tune" it to get various results depending on what they wanted it to do. They found that they could change the forward "gain" of it and also that they could change the way it performed in other aspects. A three element yagi will have about the same gain as a two element quad.

Let me say a word about "gain." Gain is a term used but not well understood. The antenna does not amplify anything. A hundred watts delivered to a dipole or a beam get you 100 watts of power out of both, (assuming 100% efficiency). The beam takes that energy and focuses it into a beam. If you were standing in front of the dipole with a field strength meter you would see only the portion of power covered in a 360 degree radiation pattern. All of the radiated energy is focused into a beam with a Yagi and thus it looks much more powerful, and it is, in one direction. Gain is measured with a dipole as a reference. There are two types of references of an antenna.

Gain is measured in bells or decibels (one-tenth of a bell). This is not a linear operation. It is logarithmic. Gain is a measure of the ability of a circuit to increase the power or amplitude of a signal from the input to the output. It is also the ability of an antenna to appear to amplify power. Gain is usually defined as the mean ratio of the signal output of a system to the signal input of the same system. An antenna does not really amplify a single. It focuses the single in a way that brings all power into a pattern. When one measures at a position from the antenna and compares the measurement to an Omni-directional antenna the single strength is much bigger. It may also be defined on a logarithmic scale, in terms of the decimal logarithm of the same ratio ("dB gain"). A gain greater than one (zero dB), that is, amplification, is the defining property of an active component or circuit, while a passive circuit will have a gain of less than one. In antenna design, antenna gain is the ratio of power received by a directional antenna to power received by an isotropic antenna.

Gain = 10 log (P out/ P in) db

Antenna gain is measured in either **dBi** or **dBd**.

It is important to note that antenna gain is different than amplifier gain. Antennas do not have a power source that allows the antenna to create additional energy to boost the signal. An antenna is similar to a reflective lens in principle - it takes the energy available from the source and focuses it over a wider or narrower area.

Antenna gain is then a measure of the amount of focus that an antenna can apply to the incoming signal relative to one of two reference dispersion patterns. Digi specifies all antenna gains in dBi.

dBi is the amount of focus applied by an antenna with respect to an "Isotropic Radiator" (a dispersion pattern that radiates the energy equally in all directions onto an imaginary sphere surrounding a point source). Thus an antenna with 2.1 dBi of gain focuses the energy so that some areas on an imaginary sphere surrounding the antenna will have 2.1 dB more signal strength than the strength of the strongest spot on the sphere around an Isotropic Radiator.

dBd refers to the antenna gain with respect to a reference dipole antenna. A reference dipole antenna is defined to have 2.15 dBi of gain. So converting between dBi and dBd is as simple as adding or subtracting 2.15 according to these formulas:

- dBi = dBd + 2.15
- dBd = dBi - 2.15

Specifying antenna gain in dBd means that the antenna in question has the ability to focus the energy x dB more than a dipole.

Beam Width

Because higher gain antennas achieve the extra power by focusing in on a smaller area it is important to remember that the greater the gain, the smaller the area covered as measured in degrees of beam width (think of an adjustable beam flashlight). In many cases a high gain antenna is a detriment to the system performance because the system needs to have reception over a large area.

THE ELEMENTS OF A YAGI

THE DRIVEN ELEMENT
The driven element of a Yagi is the feed point where the feed line is attached from the transmitter to the Yagi to perform the transfer of power from the transmitter to the antenna.
A dipole driven element will be "resonant" when its electrical length is 1/2 of the wavelength of the frequency applied to its feed point.
The feed point in the picture above is on the center of the driven element.

THE DIRECTOR
The director/s is the shortest of the parasitic elements and this end of the Yagi is aimed at the receiving station. It is resonant slightly higher in frequency than the driven

element, and its length will be about 5% shorter, progressively than the driven element. The director/s length/s can vary, depending upon the director spacing, the number of directors used in the antenna, the desired pattern, pattern bandwidth and element diameter. The number of directors that can be used are determined by the physical size (length) of the supporting boom needed by your design.

The director/s are used to provide the antenna with directional pattern and gain.

The amount of gain is directly proportional to the length of the antenna array and not by the number of directors used. The spacing of the directors can range from .1 wavelength to .5 wavelength or more and will depend largely upon the design specifications of the antenna.

THE REFLECTOR

The reflector is the element that is placed at the rear of the driven element (The dipole). It's resonant frequency is lower, and its length is approximately 5% longer than the driven element. It's length will vary depending on the spacing and the element diameter. The spacing of the reflector will be between .1 wavelength and .25 wavelength. It's spacing will depend upon the gain, bandwidth, F/B ratio, and sidelobe pattern requirements of the final antenna design.

BANDWIDTH AND IMPEDANCE

The impedance of an element is its value of pure resistance at the feed point plus any reactance (capacitive or inductive) that is present at that feed point. Of primary importance here is the impedance of the driven element, the point on the antenna where the transfer of rf from the feedline takes place.

Maximum energy transfer of rf at the design frequency occurs when the impedance of the feed point is equal to the impedance of the feedline. In most antenna designs, the feedline impedance will be 50 ohms, but usually the feed point impedance of the Yagi is rarely 50 ohms. In most cases it can vary from approximately 40 ohms to around 10 ohms, depending upon the number of elements, their spacing and the antenna's pattern bandwidth. If the feedline impedance does not equal the feed point impedance, the driven element cannot transfer the rf energy effectively from the transmitter, thus reflecting it back to the feedline resulting in a Standing Wave Ratio. Because of this, impedance matching devices are highly recommended for getting the best antenna performance.

The impedance bandwidth of the driven element is the range of frequencies above and below the center design frequency of the antenna that the driven element's feed point will accept maximum power (rf), from the feedline.

The design goal is to have the reactance at the center design frequency of the Yagi = (0),,, (j + 0).

The impedance matching device will now operate at it's optimum bandwidth. Wide element spacing, large element diameter, wide pattern bandwidth, and low "Q" matching systems will all add to a wider impedance bandwidth.

ABOUT ANTENNA PATTERNS

The antenna's radiation pattern or polar plot as it is sometimes called plays a major role in the overall performance of the Yagi antenna.

The directional gain, front-to-back ratio, beamwidth, and unwanted (or wanted) sidelobes combine to form the overall radiation pattern. The antenna's radiation pattern bandwidth is the range of frequencies above and below the design frequency in which the pattern remains consistent.

The amount of variation from the antenna's design specification goals that can be tolerated is subjective, and limits put into the design are mainly a matter of choice of the designer. "In other words.....trade offs".

Equal spaced, equal length directors may give higher gain at a particular frequency, but the bandwidth is more narrow and larger sidelobe levels are created.

Wide spacing will increase the bandwidth, but the sidelobes become large.

By varying both the spacing and director lengths the pattern and the pattern bandwidth may be more controlled.

More directors within a given boom length won't increase the gain by any great amount, but will give you better control of the antenna's pattern over a wider range of frequencies in the band of design.

If you reduce the length of each succeeding director by a set factor (%), AND increase the spacing of each succeeding director by another factor, a very clean pattern with good pattern bandwidth can be obtained.

The TRADE OFF......will be a small loss in the optimum forward gain (10% to 15%).

In a nutshell......when you make a change to one part of the antenna, this changes the performance of another part.....all changes interact with each other and the final performance!

GAIN vs FRONT-TO-BACK RATIO

With highest forward gain design, the main lobe becomes narrower in both the elevation and azimuth planes, and a backlobe is always present. When you design "out" the backlobe, the pattern gets wider and the forward gain goes down. In some cases, the sidelobes become quite large.

The Balanced Feed and Unbalanced Feed.

The Balanced feed system:
This may give a broader impedance bandwidth, but the main problem is that the driven element must in most cases be split in the center and insulated from the boom. Construction considerations aside, it is the better of the feed systems. Meeting the requirements of a balanced matching system is usually the main problem, but there are many methods available.

One method is to not split the driven element and use a "T" match, which can be described as two gamma matches on each side of the center of the element, fed with a 1:1 balun at the center.

The main drawback is that it's difficult to adjust.

The Unbalanced feed system:
Another method (for low impedance feed points) uses a split element insulated from the boom, and is fed with a "down-step 4:1 balun" made by combining two 1/4 wavelength

sections of coaxial feedline in parallel, attaching an equal length of insulated wire to the outside of these sections, and connecting it to the center conductors at the feed point end and to the shields at the feed-line end. The impedance of this type of "balun" should be at or near the mid-point value between the feed point impedance and the feedline impedance.
For example, two 75 ohm sections paralleled will equal 37.5 ohms and will match a 25 ohm feed point to a 50 ohm feedline with a 1.0 to 1 SWR.

The most common method in use by hams today is the gamma match. It will provide an easy and sure method of matching to the feed point without any loss of bandwidth.

A Cubical Quad antenna is made up of two or more loops, usually square, measuring about one wavelength in length.

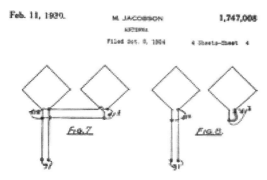

A two element Quad has about the same gain as a 3 element Yagi.

The Complete Study Guide For All Amateur Radio Tests

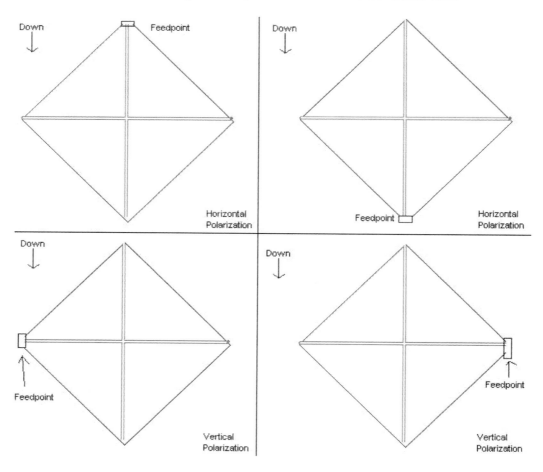

The Complete Study Guide For All Amateur Radio Tests

Quad Loop with Supporting structure shown

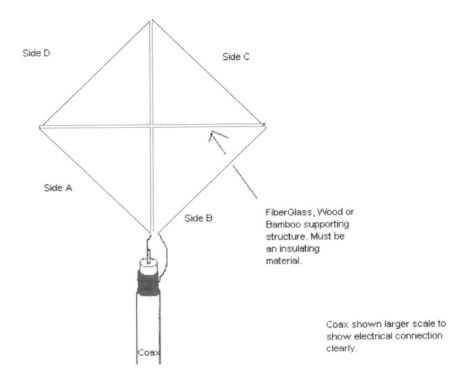

Electrical Makeup of Quad Loop

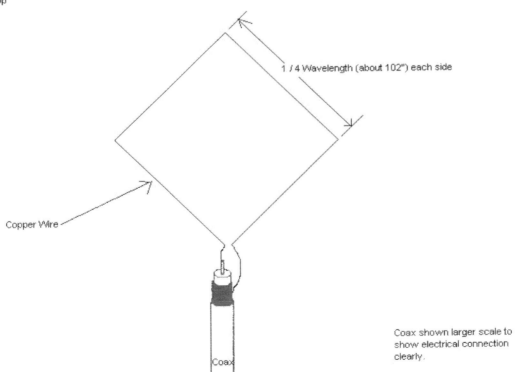

Beverage antenna information

The Beverage (or "wave") antenna was invented in the early 1920s by Dr. Harold H. Beverage. It was first discussed in a paper titled "The Wave Antenna - A New Type of Highly Directive Antenna" written by Beverage, Chester W. Rice and Edward W. Kellogg for the journal of the American Institute of Electrical Engineers (Volume 42, 1923). The paper discusses testing longwave antennas (7,000 to 25,000 meters; 12-43 kHz) that were 7 miles (11 km) long. This work was done at Riverhead, Long Island, NY, and mentions "shortwave" tests around 450 meters (665 kHz) as a practical upper limit in subsequent experiments. While others have since written about the antenna, if you can find a reprint of this original work in a research library, you'll find the paper is fascinating reading.

In 1938, the Radio Institute of America presented Dr. Beverage with its Armstrong Medal for his work in the development of antenna systems. The Beverage antenna, the citation said, was "the precursor of wave antennas of all types." Dr. Harold Henry Beverage, Stony Brook, NY, USA, passed away on January 27, 1993 (at age 99).

A classic Beverage receiving antenna requires a lot of space. It is a long wire, one or more wavelengths long, mounted near to the ground and oriented in the direction of the desired reception. A nominal 9:1 balun is required at the juncture of the wire and 50- or 75-Ohm coaxial feedline. The far end is terminated with a nominal 600-ohm resistance. (When available land will not permit the installation of a "full length" Beverage, some people install "short" Beverages, ranging in length from about 300 feet up to 600 feet or so.)

The Beverage antenna is highly directional, responsive to low-angle signals, has little noise pick-up, and produces excellent signal to noise ratios.

Folded Dipoles

A variation of the dipole is an antenna called a folded dipole. It radiates like a dipole but sort of looks like a **squashed** quad.

Having a folded dipole does **not** mean that you have an antenna that is folded in half and so you obtain an antenna that now takes up half the space of a regular dipole. No, the antenna is still approximately the same length as a regular dipole. It *is* however, an antenna that has a wire folded back over itself, hence its name. Below is a picture of a folded dipole.

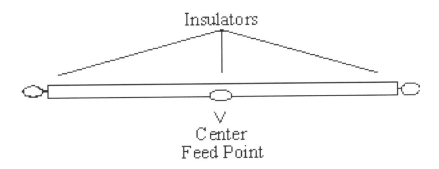

  The starting formula for the folded dipole calculation is the same as a dipole, 468 / Frequency (in MHz). Let's try an example: Design a folded dipole for the 40 meter band. The frequency that is chosen might be 7.15 MHz. Plugging this in to the formula (468 / 7.15) gives a folded dipole with a length of 65.45 feet. When I modeled a dipole on the computer at 30 feet, I came up with a length of 65.47 feet. When I added a second wire to make the folded dipole shown above, I designed the antenna with 1 inch spacing between the two wires. Note that this adds 1 more inch to each of the two antenna wires over that of a single wire dipole. This plus the fact that we are actually turning up the ends of the antenna, means that the horizontal length actually need to be a little shorter to be once again at resonance. The total length came to 64.38 feet, 1.09 feet shorter than the straight dipole. If you use a greater spacing, say 1 foot between the wires the length is 63.1 feet, 2.37 feet shorter. So be sure to shorten the antenna a bit or you'll find yourself operating lower down the band than you expected.

The feed point impedance is also modified by the second wire. Let's say the original dipole was 72 ohms. The step-up for a two wire folded dipole is 4 times which means 4 * 72 = ~288 ohms. (The computer shows 281 ohms on my example, but remember, we reduced the length slightly also.) This step up continues if you add more and more wires. A three wire antenna would provide a step-up of 9, and a four wire antenna provides a step-up of 16.

We can see why this step-up occurs by looking at the power formula P=(I*I) * R, this can be rewritten as R = P / (I*I). If the power to a regular dipole antenna was 100 watts and the current was 1.2 amps, we'd solve for R as R = 100 / (1.2*1.2), which is the same as R = 100 / 1.44, which is 69.44 ohms. In the folded dipole the wires are in parallel, the current must be divided between the two wires. The current in each is half and the total power has not changed, so now the formula is R = 100 / (.6*.6), which is the same as R = 100 / .36, which of course is 277.77 ohms, 4 times the normal dipole antenna.

  Antennas must be matched to the feed line and the feed line must be matched to the radio. The impendence must match throughout the system for the best and maximum transfer of power to the antenna and out. There are several ways to achieve a match.

All antennas have directional qualities. They do not radiate power equally in all directions. Therefore, antenna radiation patterns or plots are a very important tool to both the antenna designer and the end user. These plots show a quick picture of the overall antenna response. However, radiation patterns can be confusing. Each antenna supplier/user has different standards as well as plotting formats. Each format has its own pluses and minuses. Hopefully this technical note will shed some light on understanding and using antenna radiation patterns.

This figure shows a rectangular azimuth ("E" plane) plot presentation of a typical 10 element Yagi. The detail is good but the pattern shape is not always apparent

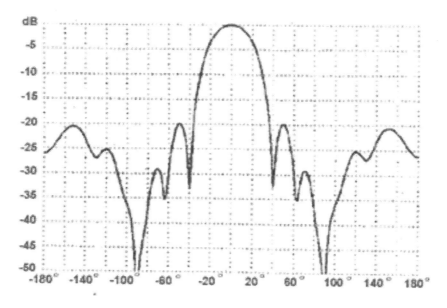

Antenna Radiation Patterns:

Antenna radiation plots can be quite complex because in the real world they are three-dimensional. However, to simplify them a Cartesian coordinate system (a two-dimensional system which refers to points in free space) is often used. Radiation plots are most often shown in either the plane of the axis of the antenna or the plane perpendicular to the axis and are referred to as the **azimuth or "E-plane" and the elevation or "H-plane"** respectively.

Many plotting formats or grids are in use. Rectangular grids (Figure 1) as well as polar coordinate systems (Figure 2) are in wide use. The principal objective is to show a radiation plot that is representative of a complete 360 degrees
in either the azimuth or the elevation plane. In the case of highly directional antennas, the radiation pattern is similar to a flashlight beam.

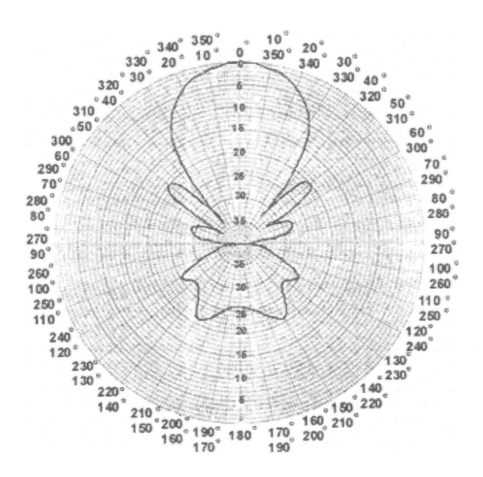

This is a polar plot of the same 10 element Yagi and is similar to a compass rose. Therefore it is more compatible with maps and directions. Note that it shows the sidelobes of the antenna relative to the main beam in decibels. This type of plot is preferred when the exact level of the sidelobes is important.

In the VHF/UHF and microwave region, the antenna radiation plot shows the relative field intensity in the far field (at least 100 feet or 30 meters distant from typical antennas) in free space at a distant point. Ground reflections are usually not a factor at these frequencies so they are often ignored. The antenna supplier either measures the radiation pattern by rotating the antenna on its axis or calculates the signal strength around the points of the compass with respect to the main beam peak. This provides a quick reference to the response of the antenna in any direction. Note that the antenna radiation pattern is reciprocal so it receives and transmits signals in the same direction.

For ease in use, clarity and maximum versatility, radiation plots are usually normalized to the outer edge of the coordinate system. Furthermore, most of us are not accustomed to thinking in terms of signal strength in volts, microvolts etc. so radiation plots are usually shown in relative dB (decibels).

For those not familiar with decibels, they are used to express differences in power in a logarithmic fashion. A drop of 1 dB means that the power is decreased to about 80% of the original value while a 3 dB drop is a power decrease of 50% or one-half the power.

The beamwidth specified on most data sheets is usually the 3 dB or half-power beamwidth.

A 10 dB drop is considered a large drop, a decrease to 10% of the original power level.

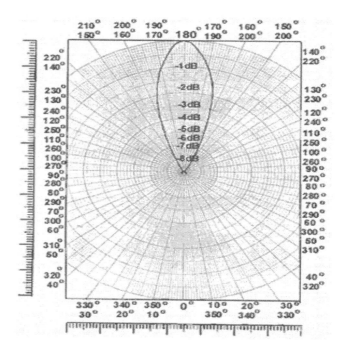

This is a linear plot of a yagi antenna

Another reason for using dB is that successive dB can be easily added or subtracted. A doubling of power is 3 dB while a quadrupling is 6 dB. Therefore, if the antenna gain is doubled (3 dB) and the transmitter power is quadrupled (6 dB), the overall improvement is 9 dB. Likewise, dB can also be subtracted.

Three types of plotting scales are in common usage; linear, linear logarithmic and modified logarithmic. The linear scale (Figure 3) emphasizes the main radiation beam. Some would argue that this plotting system makes the radiation pattern look better than it really is since it suppresses all side lobes. The linear logarithmic scale (figure 2) is preferred when the level of all sidelobes is important. The modified logarithmic scale (figure 4) emphasizes the shape of the major beam while compressing very low-level (>30 dB) sidelobes towards the center of the pattern. This plotting scale is now becoming quite popular.

How to interpret antenna radiation plots?
An antenna plot is like a road map. It tells you where the radiation is concentrated. Patterns are usually referenced to the outer edge of the plot which is the maximum gain of the antenna. This makes it easy to determine other important antenna characteristics directly from the plot.

Most antenna users are interested in the directivity or beamwidth of the antenna. This is usually referred to as the "half-power" or 3 dB beamwidth, the points between which half the power is radiated or concentrated, and specified in degrees. As an example, the typical half-power beamwidths of a 3, 6 and 10 element Yagi are 60, 40 and 30 degrees respectively.

Another popular antenna specification is the "front-to-back" (F/B) ratio. It is defined as the difference in dB between the maximum gain or front of the antenna (usually 0 degrees) and a point exactly 180 degrees behind the front. The problem with specifying only the F/B ratio is that it does not account for any lobes in the rear two quadrants. Figure 4 shows such an example where the F/B ratio is near 26 dB but +/-30 degrees are lobes that are only 21 dB down.

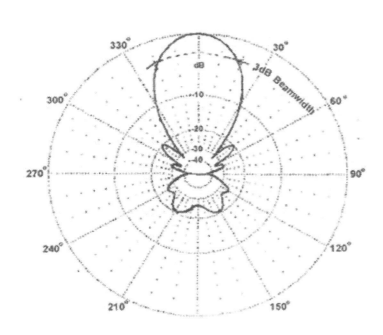

This is a modified logarithmic plot of the same 10 element Yagi which emphasizes the shape of the major beam while compressing very low-level (>30 dB) sidelobes towards the center of the pattern.

Another important antenna parameter is the side and rear lobe levels (if any). In a well designed antenna they should typically be 10-15 dB below the main beam. This parameter is often important but seldom seen on data sheets. A good logarithmic plot will easily show such lobes and the direction where they are maximum.

(No matter how good an antenna is it cannot stop waves from being distorted, such as the flutter of a signal passing through the aurora or the waving sound of HF scatter.)

How to use antenna radiation plots?

Antenna plots are the road map for the antenna user. Plots tell you where power is being radiated or received (since they are reciprocal). They also tell you how much degradation you can expect if the antenna is not aimed properly. Sometimes it is desirable to communicate with more than one station. Antenna plots will assist in the proper aiming of the antenna for optimum performance on all the desired signals. The narrower the beamwidth, the greater the difficulty in properly aiming the antenna. Remember that weather phenomenon such as wind may also affect antenna performance or dictate the type of antenna mounting.

If there are interfering signals, they may be picked up by the antenna. When you have a radiation plot, you can determine the actual level of such signals. Finally, if there are interfering signals, the radiation plot can be used to minimize them by placing such signals in a null or low sidelobe position.

Antenna radiation plots are an important tool for antenna designers and users alike. Knowing how to use them will go a long way to compare different antennas and alternative solutions. When the user has this type of data at his/her disposal, the antenna performance can be better optimized to the applications.

It takes more than a good antenna to talk to other states or nations. The ionosphere must cooperate. Solar storms have a direct effect.

In 1610 Galileo looked through his telescope and discovered the sun had spots on it. By 1745 it was noticed that the number of spots has reached a very low number, and the idea of a sunspot cycle was born. The cycle is approximately 11.5 years. A number is assigned to sunspots by looking at the number of eruption clusters and then the number of spots within those clusters.

The "sunspot number" is then given by the sum of the number of individual sunspots and ten times the number of groups. Since most sunspot groups have, on average, about ten spots. There are actually at least two "official" sunspot numbers reported. The International Sunspot Number is compiled by the Solar Influences Data Analysis Center in Belgium. The NOAA sunspot number is compiled by the US National Oceanic and Atmospheric Administration.

From 1610 to 1645 the number of sunspots were within a certain range but from about 1645 to 1715 the number shrank. Although the observations were not as extensive as in later years, the Sun was in fact well observed during this time and this lack of sunspots is well documented. This period of solar inactivity also corresponds to a climatic period called the "Little Ice Age" when rivers that are normally ice-free froze and snow fields remained year-round at lower altitudes. This period of low activity has happened in the past. It is called the Maunder Minimum. It appears the sunspots, which are eruptions of mass coronal ejections not only energize our ionosphere but may also affect weather.

The amount of energy received from the sun is measured daily in terms of the solar flux. The solar flux can vary from as low as 50 to as high as 300. During a sunspot maximum, solar flux values will typically exceed 200 resulting in excellent long distance HF communications on the 20 through 10 meter amateur bands. Solar flux values will range from 50 to 80 during sunspot minimums yielding poor long distance communications with 40 meters (7 MHz) typically being the highest usable frequency band.

An increase in solar flux values for a period of several days generally indicates an improvement in long distance HF communications during that time period. For example, the highest usable frequency will generally increase and HF communications improve if the solar flux has been running about 110 and then jumps to around 130 for several days. In contrast, the highest usable frequency will decrease and HF communications deteriorate if the solar flux instead falls to 90.

Solar Flux	Expected Condition
50 – 70	Above 40 meters is unstable
70 – 90	20 meters and below is poor
90 – 120	15 meters and below is fair
Over 200	6 meters and below is good

The sun is continuously ejecting large quantities of changed particles (atoms stripped of their electrons) into space. Some of these particles eventually arrive at the Earth and interact with the Earth's geomagnetic field. The amount of charged particles ejected by the sun varies from day to day and also with the 11 year sunspot cycle. The amount of particles arriving from the sun increases as the cycle approaches the sunspot maximum. Small numbers of particles arriving from the sun have relatively little affect on the Earth's geomagnetic field. Under these conditions the geomagnetic field is considered to be quite. Large numbers of charged particles can cause considerable disturbances in the geomagnetic field. A disturbed geomagnetic field is called a geomagnetic storm. It takes 8 minutes for light and x-ray energy from the sun to reach us from the sun.

For any given solar flux value, HF communications will improve when the geomagnetic field is quiet, and worsen during a geomagnetic storm. A geomagnetic storm causes the F layer to become unstable, fragment, and even seem to disappear. Storm conditions are more severe in the regions around the Earth's magnet poles since the charged particles from the sun are drawn to the poles by the Earth's magnetic field. As a result, signal paths that traverse the polar regions will be more affected by a geomagnetic storm than signal paths that cross the equator.

The condition of the geomagnetic field is measured in terms of A and K values in accordance with the following table:

A	K	Geomagnetic Field
0-3	0	Quiet
4-6	1	Quiet to unsettled
7-14	2	Unsettled
15-47	3-4	Active
48-79	5	Minor Storm
80-131	6	Major storm
132-207	7	Severe storm
208-400	8-9	Major Severe Storm

Mid latitude solar indices (solar flux, A, and K values) are broadcast at 20

The occurrences of solar flares also increases with increasing sunspot activity. A solar flare creates a burst of additional EUV energy and also ejects large quantities of charged particles into space. The EUV energy reaches the Earth in about 8 minutes creating what is know as a Sudden Ionospheric Disturbance (SID). The burst of EUV increases the ionization levels in the D, E, and F layers. The increased F layer ionization may help the propagation of high frequency signals (15 meters and above). However, the increased ionization in the D and E levels may result in the complete absorption of radio signals in the 160 through 40 meter bands and seriously degrade propagation at 30 and 20 meters. A SID may last from a few minutes to several hours, with conditions gradually returning to normal. The charged particles from the flare will arrive at the Earth in 20 to 40 hours. The particles will generally create a geomagnetic storm on their arrival. Improved HF band conditions are thus indicated by higher than normal solar flux values and low A and K values. The highest usable frequency for the skip effect propagating off of the ionosphere is called the MUF (Maximum Usable Frequency).

Mid latitude solar indices (solar flux, A, and K values) are broadcast at 20 minutes after the hour by radio station WWV on 5, 10, 15, and 20 MHz. They are also available on the Internet at www.qrz.com and in the K7VVV Solar Updates that are posted regularly on the ARRLWeb at www.arrl.org. The K7VVV updates are very good and provide links to other web sites for more information on solar indices and HF propagation. A good discussion of solar indices is also provided in the September 2002 QST magazine.

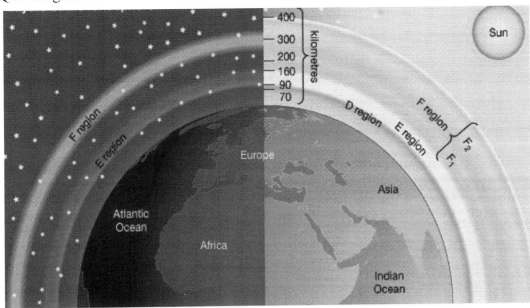

Satellite Communications

I have spent most of my life in the fields of electronics and computers. The reason for my fascination can be traced back to a single evening when I was very young. My uncle brought me outside on a clear Georgia night. He had set up a small table where sat a shortwave radio. He pointed up into the black sky and said, "Do you see that small light moving across the sky? It is a satellite. The Russians put it in space. It is called "Sputnik." Then he turned up the volume on the radio and in the distance I heard a faint "beep - beep – beep…" It was a beacon coming from the first man-made satellite, and as it streaked slowly out of sight its voice faded away. I was in love with the idea of radio and space. Later I would have a chance to be part of space once more as I worked on the Hypersonic Missile Technology project. However, it would be years later before I personally re-discovered the voices from space. Once, I heard the voice of Sputnik 1. Now, occasionally, I am heard as a voice on OSCAR 52.

One of the most exciting events in Amateur Radio was the advent of OSCAR (**OSCAR** is an acronym for **Orbiting Satellite Carrying Amateur Radio**). The first amateur satellite, simply named OSCAR 1, was launched on December 12, 1961. It was rudimentary and lasted in orbit for only 22 days. OSCAR 1 was an immediate success with over 570 amateur radio operators in 28 countries forwarding observations to Project OSCAR. These numbers are small, but even with the growth of amateur radio's population the number of those reaching for the stars are only 3000 to 5000 people world-wide.

There are several fully operational satellites in orbit acting as repeaters, beacons, "store and forward" devices, and transponders. A "store and forward device is like an orbiting email server. A person on one side of the planet can send a message to someone who does not have access to the satellite but when the satellite gets into range the receiving person can access the "bird" and retrieve the message. The satellites are projects of universities, companies, nations, and groups like the ARRL.

Currently OSCAR satellites support many different types of operation including FM voice, SSB voice, as well as digital communications of AX.25 FSK (Packet radio) and PSK-31.

Modes

The term "Mode" when applied to satellites, is not the type of operation, but it is the band or frequencies of the operating satellite. Historically OSCAR uplink (transmit to) and downlink (receive from) frequencies were designated using single letter codes.

* Mode A: 2 m uplink / 10 m downlink

* Mode B: 70 cm uplink / 2 m downlink

* Mode J: 2 m uplink / 70 cm downlink

New uplink and downlink designations use sets of paired letters following the structure X/Y where X is the uplink band and Y is the downlink band.

Designator:	H	A	V	U	L	...
Band:	15m	10m	2m	70cm	23cm	...
Frequency	H- 21MHz	A-29MHz	V-145MHz	U-435MHz	L- 1.2GHz	...

There are only a handful of satellites still in operation. Many have been launched and have died in space. Because we are no longer flush with cash we have not placed an Amateur Radio satellite in orbit for several years. The oldest satellite is AO 7. It was launched in the 70's. It failed in the 80's then miraculously came back to life. It is in limited operation today. It also has the highest orbit at around 14000 miles. This is about twice as high as the other satellites, which are considered L.E.Os (Low Orbiting Satellites), giving it a larger footprint.

Here is a list of functioning satellites.

Oscar Number: AO-27

Mode V/U (J) FM Voice Repeater: Operational

Uplink: 145.8500 MHz FM

Downlink 436.7950 MHz FM

Oscar Number: SO-50

Uplink: 145.8500 MHz FM, PL 67.0 Hz.

Downlink 436.7950 MHz FM

Oscar Number: 52

Uplink: 435.2250 - 435.2750 MHz SSB/CW

Downlink 145.9250 - 145.8750 MHz SSB/CW

Oscar Number: FO-29

Uplink: 145.9000 - 146.0000 MHz SSB/CW

Downlink 435.8000 - 435.9000 MHz SSB/CW

Oscar Number: AO-7

Uplink: 145.8500 - 145.9500 MHz SSB/CW

Downlink 29.4000 - 29.5000 MHz SSB/CW

Or

Uplink: 432.1250 - 432.1750 MHz SSB/CW

Downlink: 145.9750 - 145.9250 MHz SSB/CW

In mid 1981 AO-7 ceased operation due to battery failures. It was thought at that time that the batteries had shorted. However on June 21, 2002, at least one of the shorted batteries went open-circuit, allowing the satellite to waken whenever it is in sunlight, and randomly begin operation in one of 4 modes.

According to the log this old warrior is still supporting transponder action in mode A and mode B as recently as 9/10/09, and on a regular basis, whenever it is in sunlight.

When the satellite is in sunlight for extended periods of time, the 24-hour timer still switches the bird between modes A and B. Listen for the corresponding beacon to determine which mode the satellite is currently operating in, or refer to the abovementioned web page to see what mode has recently supported QSO\'s as the best estimate of what is the current mode of operation.

Please remember, there are no (functional) batteries, so the satellite\'s power input is limited to whatever output can be generated by the ancient solar panels. Use the least uplink power possible to minimize your downlink power usage, and maximize the number of simultaneous QSO\'s supported in the passband. Linear transponder birds are a scarce commodity these days, so please use AO-7 responsibly, but please DO enjoy her!

Detailed Description

Although there is no rule, there is a gentleman's agreement that on SSB satellites the uplink (transmit to satellite) is on Lower Side Band and the (receive from satellite) downlink is on Upper Side Band.

General Considerations

Most Amateur Radio satellites use the U/V or V/U modes. This means an antenna that can perform on both bands or two antennas for these bands (2 Meter and 70 Centimeter) will be required. Furthermore, because the satellites are spinning or tumbling there is need of an antenna that has circular polarization or some type of arrangement that can receive both horizontally and vertically polarized signals. The satellite is moving, so the antennas should move to track the satellite, or it should be Omni directional.

The radio must have a good receiver. It is easier for the satellite to hear you than for you to hear it, since many of them use one watt or less as their transmitting power. To get the signal to and from the radio and antenna with minimal loss a good type of coaxial cable (co-ax) should be used. The connectors are also important. The common PL-259, used on most radios are not a good connector for the UHF (435MHz) band. Each PL-259 will introduce about .3 to .5 db. of loss. That is not very much, until you start adding all of the connectors together. The "N" type connector is best for UHF.

As my first satellite radio I chose a Kenwood 2000X. This turned out to be a great mistake and disappointment. Some of the best satellites downlink on a frequency of 436.795 MHz. It just so happens that the inherent design of the Kenwood 2000X generates a "birdie" (an internally produced oscillation) on that frequency, which completely drowns out all incoming signals, making the radio useless for its intended purpose. There are a few satellites that use different frequencies and the 2000X works for those birds quite well. Learn from my error and do your homework by talking to people who have used the rig for the purpose you are wishing to use it. Do not believe the company advertisements.

Generally speaking, the rig you want is a radio that will transmit and receive in the 2 Meter and 70 Centimeter bands on C.W., F.M., S.S.B., and will handle the digital modes you wish to use.

My Favorite satellite to work is OSCAR 52.

VO-52, or OSCAR 52 is a microsatellite working in the V/U mode.

Its antennas are a Turnstile design for both the UHF receiver and the VHF Transmitter. The unit is designed to have an inverted transmit and receive frequency coordination. This means there is a range the satellite transmits and receives in. The center frequency of the transponders are:

- Transponder Uplink: 435.250 MHz
- Transponder Downlink: 145.900 MHz

But if I transmit to the uplink on 435.255 the satellite will transmit back to me on 145.895. As I move higher it will move lower and if I transmit on Upper Side Band it will transmit on Lower Side Band. If I transmit to it on Lower Side Band it will transmit back on Upper Side band. It does the opposite of what I do. Most of the SSB birds are like this. They have inverted transponders.

Needless to say, this can get very confusing if you do not realize what is happening when you begin to listen for your voice to come back. You must align your transmit and receive frequencies or you will not hear the person attempting to answer you. After you can hear yourself clearly through the satellite you are ready for your first contact from space. OSCAR 52 is a good place to start.

VO-52 is India's contribution to the international community of Amateur Radio Operators. This effort is also meant to bring ISRO's satellite services within the reach of the common man and popularize space technology among the masses.

This satellite will play a valuable role in the national and international scenario by providing a low cost readily accessible and reliable means of communications during emergencies and calamities like floods, earthquakes etc. It will stimulate technical interest and awareness among the younger generation by providing them with an opportunity to develop their technological projects including offering a platform. Some of the new technologies being tested in VO-52 include Integrated Processor based Electronic Bus Management Unit, Lithium Ion Battery and Gallium Arsenide based Solar Panels.

Satellite Features:

- Physical: 630 mm X 630 mm X 550 mm Cuboid
- Mass: 42.5 kg
- Orbit: Near Circular Polar Low Earth Orbit
- Structure: Aluminium Honeycomb Structure
- Power: Body mounted Gallium Arsenide Solar Panels, Lithium Ion Battery
- Stabilisation: Spin Stabilized (4 +/- 0.5 RPM)
- Antennas: UHF Turnstile VHF Turnstile
- Transponder Uplink: 435.250 MHz
- Transponder Downlink: 145.900 MHz
- Beacons - 145.936 MHz (Unmodulated Carrier) OR 145.860 MHz CW Telemetry
- Transponder Bandwidth: 60 kHz
- Transmitter Output: 1 watt

Modes of Communication: CW, SSB

Yes, it is a SSB "INVERTING" transponder. This means if you transmit in the LSB the satellite will relay it down in the USB and if the uplinked frequency goes up the downlink frequency will go down. It can be very confusing. If you are not prepared you and your contact can chase each other across the band.

Doppler shift on SSB and CW can be a real issue. Due to the high orbital speed of the OSCAR satellites, the uplink and downlink frequencies will vary during the course of a satellite pass. This phenomenon is known as the Doppler effect. While the satellite is moving towards the ground station, the downlink frequency will appear to be higher than normal and therefore, the receiver frequency at the ground station must be adjusted higher in order to continue receiving the satellite. With SSB communications this will be very noticeable in that the actual frequency of the voices will rise and fall. The satellite in turn, will be receiving the uplink signal at a higher frequency than normal so the ground station's transmitted uplink frequency must be lower in order to be received by the satellite. After the satellite passes overhead and begins to move away, this process reverses itself. The downlink frequency will appear lower and the uplink frequency will need to be adjusted higher. The following mathematical formulas relate the Doppler shift to the velocity of the satellite. Since the bird is moving rather quickly it is best to have an antenna that moves also. Although Omni directional antennas can work, they are not nearly as good as a set of beams. I tried a set of Eggbeaters (called this because they consist of two rings at 90 degrees) but I was disappointed in the performance as compared to a beam.

My first antenna of choice is a set of Arrow dual bands co-phased together. They were stationary. They had stationary elevation but could rotate. Elements allow for vertically and horizontally polarized signals from both bands to be received and transmitted at once. This eliminates fading as the satellite spins.

The antennas are tilted about 30 degrees skyward to catch the satellites as they fly by. If a person really wanted to do this project right they would install a second rotator to raise and lower the angle of the antennas, but this will have to do for now. Because there must be two antennas (one horizontal and one vertical) on each band some phasing had to be considered. The 2 Meter (140 Mhz) portions of each beam were co-phased and the 70 CM (440MHz) portions of the antennas were co-phased. Odd multiples of one-quarter wavelength spans of co-ax were used. The lengths should be the same and the velocity factor of the co-ax must be taken into consideration. We do not want to hit the high peak of one wave and the low peak of another wave. The result would the zero signal at the radio.

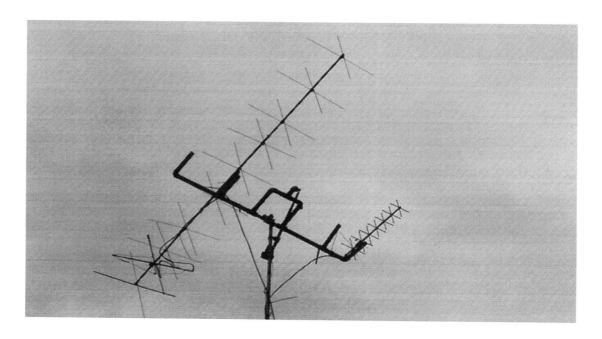

Above is the upgrade from the stationary antennas. With the addition of a home-brew elevation device these antennas are able to rotate and elevate their beams in sync. The larger of the two is a two meter ten element beam, which was purchased as a single polarity. Elements were added, along with another driven element and reflector to make it a cross (horizontal and vertical) polarized beam. The two driven elements were co-phased by cutting the coax to one-quarter wavelength with the velocity factor of the coax taken into consideration. With less than 50 watts output the satellites can be worked from AOS (acquisition of signal) to LOS (loss of signal). That is in a perfect world from the time the bird comes over the horizon to the time it drops below the horizon. HOWEVER – It should be noted (because it is mentioned on the test, That radio waves can travel slightly below the horizon. Just as red light is bent easier than blue, which gives us those beautiful red and pink sunrises and sunsets, so radio waves bend easier, allowing the bent waves to travel farther than the eye can see from horizon to horizon and slightly below it.

There are radios that have only one antenna input for both 440 MHz and 140 MHz. For those radios we can phase the two frequencies together by using three-quarter wavelength on the 440 MHz band and one-quarter wavelength on the 140 MHz band. The result will be close enough to work.

I cannot explain what a thrill it was when I heard my own voice coming back to me from space. That first QSO with a gent in Iowa was great as we both tracked the satellite toward Canada.

For a list of satellites and the information needed to track them, go to www.amsat.org (Amateur Radio Satellite Corporation).

Mores Code –

Although code is no longer needed on the tests I have included the specifics here.

Oddly, Morse Code can be heard everyday even in satellite communications and moon bounce. It is reliable and can be detected where the subtleties of the human voice could not be discerned.

Letter	Morse
A	.-
B	-...
C	-.-.
D	-..
E	.
F	..-.
G	--.
H	....
I	..
J	.---

The Complete Study Guide For All Amateur Radio Tests

K	-.-
L	.-..
M	--
N	-.
O	---
P	.--.
Q	--.-
R	.-.
S	...
T	-
U	..-
V	...-
W	.--
X	-..-
Y	-.--
Z	--..

Ä	.-.-
Á	.--.-
Å	.--.-
Ch	----
É	..-..
Ñ	--.--
Ö	---.
Ü	..--

Digit	Morse
0	-----
1	.----
2	..---
3	...--
4	....-
5	.....
6	-....
7	--...
8	---..

9	---- .

Punctuation Mark	Morse
Full-stop (period)	.-.-.-
Comma	--..--
Colon	---...
Question mark (query)	..--..
Apostrophe	.----.
Hyphen	-....-
Slash ("/")	-..-.
Brackets (parentheses)	-.--.-
Quotation marks	.-..-.
At sign	.--.-.
Equals sign	-...-

The prosigns are combinations of two letters sent together with no space in between. The other abbreviations and Q codes are sent with the normal spacing.

Prosign	Morse
AR, End of message	.-.-.
AS, Wait	.-...

BT (or TV), Break in the text	-...-
CL, Going off the air ("clear")	-.-..-..
SK, End of transmission	...-.-

Other Phrases	Abbreviation
Over	K
Roger	R
See you later	CUL
Be seeing you	BCNU
You're	UR
Signal report	RST
Best regards	73

Notes

If the duration of a dot is taken to be one unit then that of a dash is three units. The space between the components of one character is one unit, between characters is three units and between words seven units. To indicate that a mistake has been made and for the receiver to delete the last word, send .... (eight dots).

Rules and Regulations
FCC part 97

Title 47 – Part 97 Subpart A—General Provisions § 97.1 Basis and purpose.

The rules and regulations in this part are designed to provide an amateur radio service having a fundamental purpose as expressed in the following principles:

(a) Recognition and enhancement of the value of the amateur service to the public as a voluntary noncommercial communication service, particularly with respect to providing emergency communications.

(b) Continuation and extension of the amateur's proven ability to contribute to the advancement of the radio art.

(c) Encouragement and improvement of the amateur service through rules which provide for advancing skills in both the communication and technical phases of the art.

(d) Expansion of the existing reservoir within the amateur radio service of trained operators, technicians, and electronics experts.

(e) Continuation and extension of the amateur's unique ability to enhance international goodwill.

§ 97.3 Definitions.

(a) The definitions of terms used in part **97** are:

(1) *Amateur operator.* A person named in an amateur operator/primary license station grant on the ULS consolidated licensee database to be the control operator of an amateur station.

(2) *Amateur radio services.* The amateur service, the amateur-satellite service and the radio amateur civil emergency service.

(4) *Amateur service.* A radiocommunication service for the purpose of self-training, intercommunication and technical investigations carried out by amateurs, that is, duly authorized persons interested in radio technique solely with a personal aim and without pecuniary interest.

(5) *Amateur station.* A station in an amateur radio service consisting of the apparatus necessary for carrying on radiocommunications.

(6) *Automatic control.* The use of devices and procedures for control of a station when it is transmitting so that compliance with the FCC Rules is achieved without the control operator being present at a control point.

(7) *Auxiliary station.* An amateur station, other than in a message forwarding system, that is transmitting communications point-to-point within a system of cooperating amateur stations.

(8) *Bandwidth.* The width of a frequency band outside of which the mean power of the transmitted signal is attenuated at least 26 dB below the mean power of the transmitted signal within the band.

(9) *Beacon.* An amateur station transmitting communications for the purposes of observation of propagation and reception or other related experimental activities.

(10) *Broadcasting.* Transmissions intended for reception by the general public, either direct or relayed.

(11) *Call sign system.* The method used to select a call sign for amateur station over-the-air identification purposes. The call sign systems are:

(i) *Sequential call sign system.* The call sign is selected by the FCC from an alphabetized list corresponding to the geographic region of the licensee's mailing address and operator class. The call sign is shown on the license. The FCC will issue public announcements detailing the procedures of the sequential call sign system.

(ii) *Vanity call sign system.* The call sign is selected by the FCC from a list of call signs requested by the licensee. The call sign is shown on the license. The FCC will issue public announcements detailing the procedures of the vanity call sign system.

(iii) *Special event call sign system.* The call sign is selected by the station licensee from a list of call signs shown on a common data base coordinated, maintained and disseminated by the amateur station special event call sign data base coordinators. The call sign must have the single letter prefix K, N or W, followed by a single numeral 0 through 9, followed by a single letter A through W or Y or Z (for example K1A). The special event call sign is substituted for the call sign shown on the station license grant while the station is transmitting. The FCC will issue public announcements detailing the procedures of the special event call sign system.

(12) *CEPT radio amateur license.* A license issued by a country belonging to the European Conference of Postal and Telecommunications Administrations (CEPT) that has adopted Recommendation T/R 61–01 (Nice 1985, Paris 1992, Nicosia 2003).

(13) *Control operator.* An amateur operator designated by the licensee of a station to be responsible for the transmissions from that station to assure compliance with the FCC Rules.

(14) *Control point.* The location at which the control operator function is performed.

(15) *CSCE.* Certificate of successful completion of an examination.

(16) *Earth station.* An amateur station located on, or within 50 km of, the Earth's

surface intended for communications with space stations or with other Earth stations by means of one or more other objects in space.

(17) [Reserved]

(18) *External RF power amplifier.* A device capable of increasing power output when used in conjunction with, but not an integral part of, a transmitter.

(19) [Reserved]

(20) *FAA.* Federal Aviation Administration.

(21) *FCC.* Federal Communications Commission.

(22) *Frequency coordinator.* An entity, recognized in a local or regional area by amateur operators whose stations are eligible to be auxiliary or repeater stations, that recommends transmit/receive channels and associated operating and technical parameters for such stations in order to avoid or minimize potential interference.

(23) *Harmful interference.* Interference which endangers the functioning of a radionavigation service or of other safety services or seriously degrades, obstructs or repeatedly interrupts a radiocommunication service operating in accordance with the Radio Regulations.

(24) *IARP (International Amateur Radio Permit).* A document issued pursuant to the terms of the Inter-American Convention on an International Amateur Radio Permit by a country signatory to that Convention, other than the United States. Montrouis, Haiti. AG/doc.3216/95.

(25) *Indicator.* Words, letters or numerals appended to and separated from the call sign during the station identification.

(26) *Information bulletin.* A message directed only to amateur operators consisting solely of subject matter of direct interest to the amateur service.

(27) *In-law.* A parent, stepparent, sibling, or step-sibling of a licensee's spouse; the spouse of a licensee's sibling, step-sibling, child, or stepchild; or the spouse of a licensee's spouse's sibling or step-sibling.

(28) *International Morse code.* A dot-dash code as defined in ITU–T Recommendation F.1 (March, 1998), Division B, I. Morse code.

(29) *ITU.* International Telecommunication Union.

(30) *Line A.* Begins at Aberdeen, WA, running by great circle arc to the intersection of 48° N, 120° W, thence along parallel 48° N, to the intersection of 95° W, thence by great circle arc through the southernmost point of Duluth, MN, thence by great circle arc to 45° N, 85° W, thence southward along meridian 85° W, to its intersection with parallel 41° N, thence along parallel 41° N, to its intersection with meridian 82° W, thence by great circle arc through the southernmost point of Bangor, ME, thence by great circle arc through the southernmost point of Searsport, ME, at which point it terminates.

(31) *Local control.* The use of a control operator who directly manipulates the operating adjustments in the station to achieve compliance with the FCC Rules.

(32) *Message forwarding system.* A group of amateur stations participating in a voluntary, cooperative, interactive arrangement where communications are sent from the control operator of an originating station to the control operator of one or more destination stations by one or more forwarding stations.

(33) *National Radio Quiet Zone.* The area in Maryland, Virginia and West Virginia Bounded by 39°15' N on the north, 78°30' W on the east, 37°30' N on the south and 80°30' W on the west.

(34) *Physician.* For the purpose of this part, a person who is licensed to practice in a place where the amateur service is regulated by the FCC, as either a Doctor of Medicine (M.D.) or a Doctor of Osteophathy (D.O.)

(35) *Question pool.* All current examination questions for a designated written examination element.

(36) *Question set.* A series of examination questions on a given examination selected from the question pool.

(37) *Radio Regulations.* The latest ITU *Radio Regulations* to which the United States is a party.

(38) *RACES* (radio amateur civil emergency service). A radio service using amateur stations for civil defense communications during periods of local, regional or national civil emergencies.

(39) *Remote control.* The use of a control operator who indirectly manipulates the operating adjustments in the station through a control link to achieve compliance with the FCC Rules.

(40) *Repeater.* An amateur station that simultaneously retransmits the transmission of another amateur station on a different channel or channels.

(41) *Space station.* An amateur station located more than 50 km above the Earth's surface.

(42) *Space telemetry.* A one-way transmission from a space station of measurements made from the measuring instruments in a spacecraft, including those relating to the functioning of the spacecraft.

(43) *Spurious emission.* An emission, or frequencies outside the necessary bandwidth of a transmission, the level of which may be reduced without affecting the information being transmitted.

(44) *Telecommand.* A one-way transmission to initiate, modify, or terminate functions of a device at a distance.

(45) *Telecommand station.* An amateur station that transmits communications to initiate, modify or terminate functions of a space station.

(46) *Telemetry.* A one-way transmission of measurements at a distance from the measuring instrument.

(47) *Third party communications.* A message from the control operator (first party) of an amateur station to another amateur station control operator (second party) on behalf

of another person (third party).

(48) *ULS (Universal Licensing System)*. The consolidated database, application filing system and processing system for all Wireless Telecommunications Services.

(49) *VE*. Volunteer examiner.

(50) *VEC*. Volunteer-examiner coordinator.

(b) The definitions of technical symbols used in this part are:

(1) *EHF* (extremely high frequency). The frequency range 30–300 GHz.

(2) *HF* (high frequency). The frequency range 3–30 MHz.

(3) *Hz*. Hertz.

(4) *m*. Meters.

(5) *MF* (medium frequency). The frequency range 300–3000 kHz.

(6) *PEP* (peak envelope power). The average power supplied to the antenna transmission line by a transmitter during one RF cycle at the crest of the modulation envelope taken under normal operating conditions.

(7) *RF*. Radio frequency.

(8) *SHF* (super-high frequency). The frequency range 3–30 GHz.

(9) *UHF* (ultra-high frequency). The frequency range 300–3000 MHz.

(10) *VHF* (very-high frequency). The frequency range 30–300 MHz.

(11) *W*. Watts.

(c) The following terms are used in this part to indicate emission types. Refer to §2.201 of the FCC Rules, *Emission, modulation and transmission characteristics,* for information on emission type designators.

(1) *CW*. International Morse code telegraphy emissions having designators with A, C, H, J or R as the first symbol; 1 as the second symbol; A or B as the third symbol; and emissions J2A and J2B.

(2) *Data*. Telemetry, telecommand and computer communications emissions having (i) designators with A, C, D, F, G, H, J or R as the first symbol, 1 as the second symbol, and D as the third symbol; (ii) emission J2D; and (iii) emissions A1C, F1C, F2C, J2C, and J3C having an occupied bandwidth of 500 Hz or less when transmitted on an amateur service frequency below 30 MHz. Only a digital code of a type specifically authorized in this part may be transmitted.

(3) *Image*. Facsimile and television emissions having designators with A, C, D, F, G, H, J or R as the first symbol; 1, 2 or 3 as the second symbol; C or F as the third symbol; and emissions having B as the first symbol; 7, 8 or 9 as the second symbol; W as the third symbol.

(4) *MCW*. Tone-modulated international Morse code telegraphy emissions having designators with A, C, D, F, G, H or R as the first symbol; 2 as the second symbol; A or B as the third symbol.

(5) *Phone.* Speech and other sound emissions having designators with A, C, D, F, G, H, J or R as the first symbol; 1, 2 or 3 as the second symbol; E as the third symbol. Also speech emissions having B as the first symbol; 7, 8 or 9 as the second symbol; E as the third symbol. MCW for the purpose of performing the station identification procedure, or for providing telegraphy practice interspersed with speech. Incidental tones for the purpose of selective calling or alerting or to control the level of a demodulated signal may also be considered phone.

(6) *Pulse.* Emissions having designators with K, L, M, P, Q, V or W as the first symbol; 0, 1, 2, 3, 7, 8, 9 or X as the second symbol; A, B, C, D, E, F, N, W or X as the third symbol.

(7) *RTTY.* Narrow-band direct-printing telegraphy emissions having designators with A, C, D, F, G, H, J or R as the first symbol; 1 as the second symbol; B as the third symbol; and emission J2B. Only a digital code of a type specifically authorized in this part may be transmitted.

(8) *SS.* Spread spectrum emissions using bandwidth-expansion modulation emissions having designators with A, C, D, F, G, H, J or R as the first symbol; X as the second symbol; X as the third symbol.

(9) *Test.* Emissions containing no information having the designators with N as the third symbol. Test does not include pulse emissions with no information or modulation unless pulse emissions are also authorized in the frequency band.

§ 97.5 Station license required.

(a) The station apparatus must be under the physical control of a person named in an amateur station license grant on the ULS consolidated license database or a person authorized for alien reciprocal operation by §**97**.107 of this part, before the station may transmit on any amateur service frequency from any place that is:

(1) Within 50 km of the Earth's surface and at a place where the amateur service is regulated by the FCC;

(2) Within 50 km of the Earth's surface and aboard any vessel or craft that is documented or registered in the United States; or

(3) More than 50 km above the Earth's surface aboard any craft that is documented or registered in the United States.

(b) The types of station license grants are:

(1) *An operator/primary station license grant.* One, but only one, operator/primary station license grant may be held by any one person. The primary station license is granted together with the amateur operator license. Except for a representative of a foreign government, any person who qualifies by examination is eligible to apply for an operator/primary station license grant.

(2) *A club station license grant.* A club station license grant may be held only by the

person who is the license trustee designated by an officer of the club. The trustee must be a person who holds an operator/primary station license grant. The club must be composed of at least four persons and must have a name, a document of organization, management, and a primary purpose devoted to amateur service activities consistent with this part.

(3) *A military recreation station license grant.* A military recreation station license grant may be held only by the person who is the license custodian designated by the official in charge of the United States military recreational premises where the station is situated. The person must not be a representative of a foreign government. The person need not hold an amateur operator license grant.

(c) The person named in the station license grant or who is authorized for alien reciprocal operation by §**97**.107 of this part may use, in accordance with the applicable rules of this part, the transmitting apparatus under the physical control of the person at places where the amateur service is regulated by the FCC.

(d) A CEPT radio-amateur license is issued to the person by the country of which the person is a citizen. The person must not:

(1) Be a resident alien or citizen of the United States, regardless of any other citizenship also held;

(2) Hold an FCC-issued amateur operator license nor reciprocal permit for alien amateur licensee;

(3) Be a prior amateur service licensee whose FCC-issued license was revoked, suspended for less than the balance of the license term and the suspension is still in effect, suspended for the balance of the license term and relicensing has not taken place, or surrendered for cancellation following notice of revocation, suspension or monetary forfeiture proceedings; or

(4) Be the subject of a cease and desist order that relates to amateur service operation and which is still in effect.

(e) An IARP is issued to the person by the country of which the person is a citizen. The person must not:

(1) Be a resident alien or citizen of the United States, regardless of any other citizenship also held;

(2) Hold an FCC-issued amateur operator license nor reciprocal permit for alien amateur licensee;

(3) Be a prior amateur service licensee whose FCC-issued license was revoked, suspended for less than the balance of the license term and the suspension is still in effect, suspended for the balance of the license term and relicensing has not taken place, or surrendered for cancellation following notice of revocation, suspension or monetary forfeiture proceedings; or

(4) Be the subject of a cease and desist order that relates to amateur service operation and which is still in effect.

§ 97.7 Control operation required.

When transmitting, each amateur station must have a control operator. The control operator must be a person:

(a) For whom an amateur operator/primary station license grant appears on the ULS consolidated licensee database, or

(b) Who is authorized for alien reciprocal operation by §97.107 of this part. **§ 97.9 Operator license grant.**

(a) The classes of amateur operator license grants are: Novice, Technician, General, Advanced, and Amateur Extra. The person named in the operator license grant is authorized to be the control operator of an amateur station with the privileges authorized to the operator class specified on the license grant.

(b) The person named in an operator license grant of Novice, Technician, General or Advanced Class, who has properly submitted to the administering VEs a FCC Form 605 document requesting examination for an operator license grant of a higher class, and who holds a CSCE indicating that the person has completed the necessary examinations within the previous 365 days, is authorized to exercise the rights and privileges of the higher operator class until final disposition of the application or until 365 days following the passing of the examination, whichever comes first.

§ 97.11 Stations aboard ships or aircraft.

(a) The installation and operation of an amateur station on a ship or aircraft must be approved by the master of the ship or pilot in command of the aircraft.

(b) The station must be separate from and independent of all other radio apparatus installed on the ship or aircraft, except a common antenna may be shared with a voluntary ship radio installation. The station's transmissions must not cause interference to any other apparatus installed on the ship or aircraft.

(c) The station must not constitute a hazard to the safety of life or property. For a station aboard an aircraft, the apparatus shall not be operated while the aircraft is operating under Instrument Flight Rules, as defined by the FAA, unless the station has been found to comply with all applicable FAA Rules.

§ 97.13 Restrictions on station location.

(a) Before placing an amateur station on land of environmental importance or that is significant in American history, architecture or culture, the licensee may be required to take certain actions prescribed by §§1.1305–1.1319 of this chapter.

(b) A station within 1600 m (1 mile) of an FCC monitoring facility must protect that facility from harmful interference. Failure to do so could result in imposition of operating restrictions upon the amateur station by a District Director pursuant to §97.121 of this part. Geographical coordinates of the facilities that require protection are listed in §0.121(c) of this chapter.

(c) Before causing or allowing an amateur station to transmit from any place where the operation of the station could cause human exposure to RF electromagnetic field levels in excess of those allowed under §1.1310 of this chapter, the licensee is required to take certain actions.

(1) The licensee must perform the routine RF environmental evaluation prescribed by §1.1307(b) of this chapter, if the power of the licensee's station exceeds the limits given in the following table:

[1] Power = PEP input to antenna except, for repeater stations only, power exclusion is based on ERP (effective radiated power).

(2) If the routine environmental evaluation indicates that the RF electromagnetic fields could exceed the limits contained in §1.1310 of this chapter in accessible areas, the licensee must take action to prevent human exposure to such RF electromagnetic fields. Further information on evaluating compliance with these limits can be found in the FCC's OET Bulletin Number 65, "Evaluating Compliance with FCC Guidelines for Human Exposure to Radiofrequency Electromagnetic Fields."

Wavelength band	Evaluation required if power1(watts) exceeds
MF	
160 m	500
HF	
80 m	500
75 m	500
40 m	500
30 m	425
20 m	225
17 m	125
15 m	100
12 m	75
10 m	50
VHF (all bands)	50

UHF	
70 cm	70
33 cm	150
23 cm	200
13 cm	250
SHF (all bands)	250
EHF (all bands)	250
Repeater stations (all bands)	*non-building-mounted antennas:* height above ground level to lowest point of antenna <10 m *and* power >500 W ERP *building-mounted antennas:* power >500 W ERP

§ 97.15 Station antenna structures.

(a) Owners of certain antenna structures more than 60.96 meters (200 feet) above ground level at the site or located near or at a public use airport must notify the Federal Aviation Administration and register with the Commission as required by part 17 of this chapter.

(b) Except as otherwise provided herein, a station antenna structure may be erected at heights and dimensions sufficient to accommodate amateur service communications. (State and local regulation of a station antenna structure must not preclude amateur service communications. Rather, it must reasonably accommodate such communications and must constitute the minimum practicable regulation to accomplish the state or local authority's legitimate purpose. *See* PRB–1, 101 FCC 2d 952 (1985) for details.)

§ 97.17 Application for new license grant.

(a) Any qualified person is eligible to apply for a new operator/primary station, club station or military recreation station license grant. No new license grant will be issued for a Novice or Advanced Class operator/primary station.

(b) Each application for a new amateur service license grant must be filed with the

FCC as follows:

(1) Each candidate for an amateur radio operator license which requires the applicant to pass one or more examination elements must present the administering VEs with all information required by the rules prior to the examination. The VEs may collect all necessary information in any manner of their choosing, including creating their own forms.

(2) For a new club or military recreation station license grant, each applicant must present all information required by the rules to an amateur radio organization having tax-exempt status under section 501(c)(3) of the Internal Revenue Code of 1986 that provides voluntary, uncompensated and unreimbursed services in providing club and military recreation station call signs (" *Club Station Call Sign Administrator* ") who must submit the information to the FCC in an electronic batch file. The Club Station Call Sign Administrator may collect the information required by these rules in any manner of their choosing, including creating their own forms. The Club Station Call Sign Administrator must retain the applicants information for at least 15 months and make it available to the FCC upon request. The FCC will issue public announcements listing the qualified organizations that have completed a pilot autogrant batch filing project and are authorized to serve as a Club Station Call Sign Administrator.

(c) No person shall obtain or attempt to obtain, or assist another person to obtain or attempt to obtain, an amateur service license grant by fraudulent means.

(d) One unique call sign will be shown on the license grant of each new primary, club and military recreation station. The call sign will be selected by the sequential call sign system. Effective February 14, 2011, no club station license grants will be issued to a licensee who is shown as the license trustee on an existing club station license grant.

§ 97.19 Application for a vanity call sign.

(a) The person named in an operator/primary station license grant or in a club station license grant is eligible to make application for modification of the license grant, or the renewal thereof, to show a call sign selected by the vanity call sign system. Effective February 14, 2011, the person named in a club station license grant that shows on the license a call sign that was selected by a trustee is not eligible for an additional vanity call sign. (The person named in a club station license grant that shows on the license a call sign that was selected by a trustee is eligible for a vanity call sign for his or her operator/primary station license grant on the same basis as any other person who holds an operator/primary station license grant.) Military recreation stations are not eligible for a vanity call sign.

(b) Each application for a modification of an operator/primary or club station license grant, or the renewal thereof, to show a call sign selected by the vanity call sign system must be filed in accordance with §1.913 of this chapter.

(c) Unassigned call signs are available to the vanity call sign system with the following exceptions:

(1) A call sign shown on an expired license grant is not available to the vanity call sign

system for 2 years following the expiration of the license.

(2) A call sign shown on a surrendered or canceled license grant (except for a license grant that is canceled pursuant to §97.31) is not available to the vanity call sign system for 2 years following the date such action is taken. (The availability of a call sign shown on a license canceled pursuant to §97.31 is governed by paragraph (c)(3) of this section.)

(i) This 2-year period does not apply to any license grant pursuant to paragraph (c)(3)(i), (ii), or (iii) of this section that is surrendered, canceled, revoked, voided, or set aside because the grantee acknowledged or the Commission determined that the grantee was not eligible for the exception. In such a case, the call sign is not available to the vanity call sign system for 30 days following the date such action is taken, or for the period for which the call sign would not have been available to the vanity call sign system pursuant to paragraphs (c)(2) or (3) of this section but for the intervening grant to the ineligible applicant, whichever is later.

(ii) An applicant to whose operator/primary station license grant, or club station license grant for which the applicant is the trustee, the call sign was previously assigned is exempt from the 2-year period set forth in paragraph (c)(2) of this section.

(3) A call sign shown on a license canceled pursuant to §97.31 of this part is not available to the vanity call sign system for 2 years following the person's death, or for 2 years following the expiration of the license grant, whichever is sooner. If, however, a license is canceled more than 2 years after the licensee's death (or within 30 days before the second anniversary of the licensee's death), the call sign is not available to the vanity call sign system for 30 days following the date such action is taken. The following applicants are exempt from this 2-year period:

(i) An applicant to whose operator/primary station license grant, or club station license grant for which the applicant is the trustee, the call sign was previously assigned; or

(ii) An applicant who is the spouse, child, grandchild, stepchild, parent, grandparent, stepparent, brother, sister, stepbrother, stepsister, aunt, uncle, niece, nephew, or in-law of the person now deceased or of any other deceased former holder of the call sign, provided that the vanity call sign requested by the applicant is from the group of call signs corresponding to the same or lower class of operator license held by the applicant as designated in the sequential call sign system; or

(iii) An applicant who is a club station license trustee acting with a written statement of consent signed by either the licensee *ante mortem* but who is now deceased, or by at least one relative as listed in paragraph (c)(3)(ii) of this section, of the person now deceased or of any other deceased former holder of the call sign, provided that the deceased former holder was a member of the club during his or her life.

(d) The vanity call sign requested by an applicant must be selected from the group of call signs corresponding to the same or lower class of operator license held by the applicant as designated in the sequential call sign system.

(1) The applicant must request that the call sign shown on the license grant be vacated

and provide a list of up to 25 call signs in order of preference. In the event that the Commission receives more than one application requesting a vanity call sign from an applicant on the same receipt day, the Commission will process only the first such application entered into the Universal Licensing System. Subsequent vanity call sign applications from that applicant with the same receipt date will not be accepted.

(2) The first assignable call sign from the applicant's list will be shown on the license grant. When none of those call signs are assignable, the call sign vacated by the applicant will be shown on the license grant.

(3) Vanity call signs will be selected from those call signs assignable at the time the application is processed by the FCC.

(4) A call sign designated under the sequential call sign system for Alaska, Hawaii, Caribbean Insular Areas, and Pacific Insular areas will be assigned only to a primary or club station whose licensee's mailing address is in the corresponding state, commonwealth, or island. This limitation does not apply to an applicant for the call sign as the spouse, child, grandchild, stepchild, parent, grandparent, stepparent, brother, sister, stepbrother, stepsister, aunt, uncle, niece, nephew, or in-law, of the former holder now deceased.

§ 97.21 Application for a modified or renewed license grant.

(a) A person holding a valid amateur station license grant:

(1) Must apply to the FCC for a modification of the license grant as necessary to show the correct mailing address, licensee name, club name, license trustee name, or license custodian name in accordance with §1.913 of this chapter. For a club or military recreation station license grant, the application must be presented in document form to a Club Station Call Sign Administrator who must submit the information thereon to the FCC in an electronic batch file. The Club Station Call Sign Administrator must retain the collected information for at least 15 months and make it available to the FCC upon request. A Club Station Call Sign Administrator shall not file with the Commission any application to modify a club station license grant that was submitted by a person other than the trustee as shown on the license grant, except an application to change the club station license trustee. An application to modify a club station license grant to change the license trustee name must be submitted to a Club Station Call Sign Administrator and must be signed by an officer of the club.

(2) May apply to the FCC for a modification of the operator/primary station license grant to show a higher operator class. Applicants must present the administering VEs with all information required by the rules prior to the examination. The VEs may collect all necessary information in any manner of their choosing, including creating their own forms.

(3) May apply to the FCC for renewal of the license grant for another term in accordance with §§1.913 and 1.949 of this chapter. Application for renewal of a Technician Plus Class operator/primary station license will be processed as an application for renewal of a Technician Class operator/primary station license.

(i) For a station license grant showing a call sign obtained through the vanity call sign system, the application must be filed in accordance with §97.19 of this part in order to have the vanity call sign reassigned to the station.

(ii) For a primary station license grant showing a call sign obtained through the sequential call sign system, and for a primary station license grant showing a call sign obtained through the vanity call sign system but whose grantee does not want to have the vanity call sign reassigned to the station, the application must be filed with the FCC in accordance with §1.913 of this chapter. When the application has been received by the FCC on or before the license expiration date, the license operating authority is continued until the final disposition of the application.

(iii) For a club station or military recreation station license grant showing a call sign obtained through the sequential call sign system, and for a club station license grant showing a call sign obtained through the vanity call sign system but whose grantee does not want to have the vanity call sign reassigned to the station, the application must be presented in document form to a Club Station Call Sign Administrator who must submit the information thereon to the FCC in an electronic batch file. The replacement call sign will be selected by the sequential call sign system. The Club Station Call Sign Administrator must retain the collected information for at least 15 months and make it available to the FCC upon request.

(b) A person whose amateur station license grant has expired may apply to the FCC for renewal of the license grant for another term during a 2 year filing grace period. The application must be received at the address specified above prior to the end of the grace period. Unless and until the license grant is renewed, no privileges in this part are conferred.

(c) Except as provided in paragraph (a)(4) of this section, a call sign obtained under the sequential or vanity call sign system will be reassigned to the station upon renewal or modification of a station license.

§ 97.23 Mailing address.

Each license grant must show the grantee's correct name and mailing address. The mailing address must be in an area where the amateur service is regulated by the FCC and where the grantee can receive mail delivery by the United States Postal Service. Revocation of the station license or suspension of the operator license may result when correspondence from the FCC is returned as undeliverable because the grantee failed to provide the correct mailing address.

§ 97.25 License term.

An amateur service license is normally granted for a 10-year term.

§ 97.27 FCC modification of station license grant.

(a) The FCC may modify a station license grant, either for a limited time or for the duration of the term thereof, if it determines:

(1) That such action will promote the public interest, convenience, and necessity; or

(2) That such action will promote fuller compliance with the provisions of the Communications Act of 1934, as amended, or of any treaty ratified by the United States.

(b) When the FCC makes such a determination, it will issue an order of modification. The order will not become final until the licensee is notified in writing of the proposed action and the grounds and reasons therefor. The licensee will be given reasonable opportunity of no less than 30 days to protest the modification; except that, where safety of life or property is involved, a shorter period of notice may be provided. Any protest by a licensee of an FCC order of modification will be handled in accordance with the provisions of 47 U.S.C. 316.

§ 97.29 Replacement license grant document.

Each grantee whose amateur station license grant document is lost, mutilated or destroyed may apply to the FCC for a replacement in accordance with §1.913 of this chapter.

§ 97.31 Cancellation on account of the licensee's death.

(a) A person may request cancellation of an operator/primary station license grant on account of the licensee's death by submitting a signed request that includes a death certificate, obituary, or Social Security Death Index data that shows the person named in the operator/primary station license grant has died. Such a request may be submitted as a pleading associated with the deceased licensee's license. *See* §1.45 of this chapter. In addition, the Commission may cancel an operator/primary station license grant if it becomes aware of the grantee's death through other means. No action will be taken during the last thirty days of the post-expiration grace period (*see* §**97**.21(b)) on a request to cancel a license due to the licensee's death.

(b) A license that is canceled due to the licensee's death is canceled as of the date of the licensee's death.

Subpart B—Station Operation Standards § 97.101 General standards.

(a) In all respects not specifically covered by FCC Rules each amateur station must be operated in accordance with good engineering and good amateur practice.

(b) Each station licensee and each control operator must cooperate in selecting transmitting channels and in making the most effective use of the amateur service frequencies. No frequency will be assigned for the exclusive use of any station.

(c) At all times and on all frequencies, each control operator must give priority to stations providing emergency communications, except to stations transmitting communications for training drills and tests in RACES.

(d) No amateur operator shall willfully or maliciously interfere with or cause interference to any radio communication or signal.

§ 97.103 Station licensee responsibilities.

(a) The station licensee is responsible for the proper operation of the station in accordance with the FCC Rules. When the control operator is a different amateur operator than the station licensee, both persons are equally responsible for proper operation of the station.

(b) The station licensee must designate the station control operator. The FCC will presume that the station licensee is also the control operator, unless documentation to the contrary is in the station records.

(c) The station licensee must make the station and the station records available for inspection upon request by an FCC representative.

§ 97.105 Control operator duties.

(a) The control operator must ensure the immediate proper operation of the station, regardless of the type of control.

(b) A station may only be operated in the manner and to the extent permitted by the privileges authorized for the class of operator license held by the control operator.

§ 97.107 Reciprocal operating authority.

↑ top

A non-citizen of the United States ("alien") holding an amateur service authorization granted by the alien's government is authorized to be the control operator of an amateur station located at places where the amateur service is regulated by the FCC, provided there is in effect a multilateral or bilateral reciprocal operating arrangement, to which the United States and the alien's government are parties, for amateur service operation on a reciprocal basis. The FCC will issue public announcements listing the countries with which the United States has such an arrangement. No citizen of the United States or person holding an FCC amateur operator/primary station license grant is eligible for the reciprocal operating authority granted by this section. The privileges granted to a control operator under this authorization are:

(a) For an amateur service license granted by the Government of Canada:

(1) The terms of the *Convention Between the United States and Canada* (TIAS No. 2508) *Relating to the Operation by Citizens of Either Country of Certain Radio Equipment or Stations in the Other Country;*

(2) The operating terms and conditions of the amateur service license issued by the Government of Canada; and

(3) The applicable rules of this part, but not to exceed the control operator privileges of an FCC-granted Amateur Extra Class operator license.

(b) For an amateur service license granted by any country, other than Canada, with which the United States has a multilateral or bilateral agreement:

(1) The terms of the agreement between the alien's government and the United States;

(2) The operating terms and conditions of the amateur service license granted by the alien's government;

(3) The applicable rules of this part, but not to exceed the control operator privileges of an FCC-granted Amateur Extra Class operator license; and

(c) At any time the FCC may, in its discretion, modify, suspend or cancel the reciprocal operating authority granted to any person by this section.

§ 97.109 Station control.

(a) Each amateur station must have at least one control point.

(b) When a station is being locally controlled, the control operator must be at the control point. Any station may be locally controlled.

(c) When a station is being remotely controlled, the control operator must be at the control point. Any station may be remotely controlled.

(d) When a station is being automatically controlled, the control operator need not be at the control point. Only stations specifically designated elsewhere in this part may be automatically controlled. Automatic control must cease upon notification by a District Director that the station is transmitting improperly or causing harmful interference to other stations. Automatic control must not be resumed without prior approval of the District Director.

§ 97.111 Authorized transmissions.

(a) An amateur station may transmit the following types of two-way communications:

(1) Transmissions necessary to exchange messages with other stations in the amateur service, except those in any country whose administration has notified the ITU that it objects to such communications. The FCC will issue public notices of current arrangements for international communications.

(2) Transmissions necessary to meet essential communication needs and to facilitate relief actions.

(3) Transmissions necessary to exchange messages with a station in another FCC-regulated service while providing emergency communications;

(4) Transmissions necessary to exchange messages with a United States government station, necessary to providing communications in RACES; and

(5) Transmissions necessary to exchange messages with a station in a service not regulated by the FCC, but authorized by the FCC to communicate with amateur stations. An amateur station may exchange messages with a participating United States military station during an Armed Forces Day Communications Test.

(b) In addition to one-way transmissions specifically authorized elsewhere in this part, an amateur station may transmit the following types of one-way communications:

(1) Brief transmissions necessary to make adjustments to the station;

(2) Brief transmissions necessary to establishing two-way communications with other stations;

(3) Telecommand;

(4) Transmissions necessary to providing emergency communications;

(5) Transmissions necessary to assisting persons learning, or improving proficiency in, the international Morse code; and

(6) Transmissions necessary to disseminate information bulletins.

(7) Transmissions of telemetry.

§ 97.113 Prohibited transmissions.

(a) No amateur station shall transmit:

(1) Communications specifically prohibited elsewhere in this part;

(2) Communications for hire or for material compensation, direct or indirect, paid or promised, except as otherwise provided in these rules;

(3) Communications in which the station licensee or control operator has a pecuniary interest, including communications on behalf of an employer, with the following exceptions:

(i) A station licensee or control station operator may participate on behalf of an employer in an emergency preparedness or disaster readiness test or drill, limited to the duration and scope of such test or drill, and operational testing immediately prior to such test or drill. Tests or drills that are not government-sponsored are limited to a total time of one hour per week; except that no more than twice in any calendar year, they may be conducted for a period not to exceed 72 hours.

(ii) An amateur operator may notify other amateur operators of the availability for sale or trade of apparatus normally used in an amateur station, provided that such activity is not conducted on a regular basis.

(iii) A control operator may accept compensation as an incident of a teaching position during periods of time when an amateur station is used by that teacher as a part of classroom instruction at an educational institution.

(iv) The control operator of a club station may accept compensation for the periods of time when the station is transmitting telegraphy practice or information bulletins, provided that the station transmits such telegraphy practice and bulletins for at least 40 hours per week; schedules operations on at least six amateur service MF and HF bands using reasonable measures to maximize coverage; where the schedule of normal operating times and frequencies is published at least 30 days in advance of the actual transmissions; and where the control operator does not accept any direct or indirect compensation for any other service as a control operator.

(4) Music using a phone emission except as specifically provided elsewhere in this section; communications intended to facilitate a criminal act; messages encoded for the purpose of obscuring their meaning, except as otherwise provided herein; obscene or indecent words or language; or false or deceptive messages, signals or identification.

(5) Communications, on a regular basis, which could reasonably be furnished alternatively through other radio services.

(b) An amateur station shall not engage in any form of broadcasting, nor may an amateur station transmit one-way communications except as specifically provided in these rules; nor shall an amateur station engage in any activity related to program production or news gathering for broadcasting purposes, except that communications directly related to the immediate safety of human life or the protection of property may be provided by amateur stations to broadcasters for dissemination to the public where no other means of communication is reasonably available before or at the time of the event.

(c) No station shall retransmit programs or signals emanating from any type of radio station other than an amateur station, except propagation and weather forecast information intended for use by the general public and originated from United States Government stations, and communications, including incidental music, originating on United States Government frequencies between a manned spacecraft and its associated Earth stations. Prior approval for manned spacecraft communications retransmissions must be obtained from the National Aeronautics and Space Administration. Such retransmissions must be for the exclusive use of amateur radio operators. Propagation, weather forecasts, and manned spacecraft communications retransmissions may not be conducted on a regular basis, but only occasionally, as an incident of normal amateur radio communications.

(d) No amateur station, except an auxiliary, repeater, or space station, may automatically retransmit the radio signals of other amateur station.

§ 97.115 Third party communications.

(a) An amateur station may transmit messages for a third party to:

(1) Any station within the jurisdiction of the United States.

(2) Any station within the jurisdiction of any foreign government when transmitting emergency or disaster relief communications and any station within the jurisdiction of any foreign government whose administration has made arrangements with the United States to allow amateur stations to be used for transmitting international communications on behalf of third parties. No station shall transmit messages for a third party to any station within the jurisdiction of any foreign government whose administration has not made such an arrangement. This prohibition does not apply to a message for any third party who is eligible to be a control operator of the station.

(b) The third party may participate in stating the message where:

(1) The control operator is present at the control point and is continuously monitoring and supervising the third party's participation; and

(2) The third party is not a prior amateur service licensee whose license was revoked or not renewed after hearing and re-licensing has not taken place; suspended for less than the balance of the license term and the suspension is still in effect; suspended for the balance of the license term and re-licensing has not taken place; or surrendered for cancellation following notice of revocation, suspension or monetary forfeiture proceedings. The third party may not be the subject of a cease and desist order which relates to amateur service operation and which is still in effect.

(c) No station may transmit third party communications while being automatically controlled except a station transmitting a RTTY or data emission.

(d) At the end of an exchange of international third party communications, the station must also transmit in the station identification procedure the call sign of the station with which a third party message was exchanged.

§ 97.117 International communications.

Transmissions to a different country, where permitted, shall be limited to communications incidental to the purposes of the amateur service and to remarks of a personal character.

§ 97.119 Station identification.

(a) Each amateur station, except a space station or telecommand station, must transmit its assigned call sign on its transmitting channel at the end of each communication, and at least every 10 minutes during a communication, for the purpose of clearly making the source of the transmissions from the station known to those receiving the transmissions. No station may transmit unidentified communications or signals, or transmit as the station call sign, any call sign not authorized to the station.

(b) The call sign must be transmitted with an emission authorized for the transmitting channel in one of the following ways:

(1) By a CW emission. When keyed by an automatic device used only for identification, the speed must not exceed 20 words per minute;

(2) By a phone emission in the English language. Use of a phonetic alphabet as an aid for correct station identification is encouraged;

(3) By a RTTY emission using a specified digital code when all or part of the communications are transmitted by a RTTY or data emission;

(4) By an image emission conforming to the applicable transmission standards, either color or monochrome, of §73.682(a) of the FCC Rules when all or part of the communications are transmitted in the same image emission

(c) One or more indicators may be included with the call sign. Each indicator must be separated from the call sign by the slant mark (/) or by any suitable word that denotes the slant mark. If an indicator is self-assigned, it must be included before, after, or both before and after, the call sign. No self-assigned indicator may conflict with any other indicator specified by the FCC Rules or with any prefix assigned to another country.

(d) When transmitting in conjunction with an event of special significance, a station may substitute for its assigned call sign a special event call sign as shown for that station for that period of time on the common data base coordinated, maintained and disseminated by the special event call sign data base coordinators. Additionally, the station must transmit its assigned call sign at least once per hour during such transmissions.

(e) When the operator license class held by the control operator exceeds that of the station licensee, an indicator consisting of the call sign assigned to the control operator's station must be included after the call sign.

(f) When the control operator is a person who is exercising the rights and privileges authorized by §97.9(b) of this part, an indicator must be included after the call sign as follows:

(1) For a control operator who has requested a license modification from Novice Class to Technical Class: KT;

(2) For a control operator who has requested a license modification from Novice or Technician to General Class: AG;

(3) For a control operator who has requested a license modification from Novice, Technician, General, or Advanced Class to Amateur Extra Class: AE.

(g) When the station is transmitting under the authority of §97.107 of this part, an indicator consisting of the appropriate letter-numeral designating the station location must be included before the call sign that was issued to the station by the country granting the license. For an amateur service license granted by the Government of Canada, however, the indicator must be included after the call sign. At least once during each intercommunication, the identification announcement must include the geographical location as nearly as possible by city and state, commonwealth or possession.

§ 97.121 Restricted operation.

(a) If the operation of an amateur station causes general interference to the reception of transmissions from stations operating in the domestic broadcast service when receivers of good engineering design, including adequate selectivity characteristics, are used to receive such transmissions, and this fact is made known to the amateur station licensee, the amateur station shall not be operated during the hours from 8 p.m. to 10:30 p.m., local time, and on Sunday for the additional period from 10:30 a.m. until 1 p.m., local time, upon the frequency or frequencies used when the interference is created.

(b) In general, such steps as may be necessary to minimize interference to stations operating in other services may be required after investigation by the FCC.

Subpart C—Special Operations

§ 97.201 Auxiliary station.

(a) Any amateur station licensed to a holder of a Technician, General, Advanced or Amateur Extra Class operator license may be an auxiliary station. A holder of a Technician, General, Advanced or Amateur Extra Class operator license may be the control operator of an auxiliary station, subject to the privileges of the class of operator license held.

(b) An auxiliary station may transmit only on the 2 m and shorter wavelength bands, except the 144.0–144.5 MHz, 145.8–146.0 MHz, 219–220 MHz, 222.00–222.15 MHz, 431–433 MHz, and 435–438 MHz segments.

(c) Where an auxiliary station causes harmful interference to another auxiliary station, the licensees are equally and fully responsible for resolving the interference unless one station's operation is recommended by a frequency coordinator and the other station's is not. In that case, the licensee of the non-coordinated auxiliary station has primary responsibilty to resolve the interference.

(d) An auxiliary station may be automatically controlled.

(e) An auxiliary station may transmit one-way communications.

§ 97.203 Beacon station.

(a) Any amateur station licensed to a holder of a Technician, General, Advanced or Amateur Extra Class operator license may be a beacon. A holder of a Technician, General, Advanced or Amateur Extra Class operator license may be the control operator of a beacon, subject to the privileges of the class of operator license held.

(b) A beacon must not concurrently transmit on more than 1 channel in the same amateur service frequency band, from the same station location.

(c) The transmitter power of a beacon must not exceed 100 W.

(d) A beacon may be automatically controlled while it is transmitting on the 28.20–28.30 MHz, 50.06–50.08 MHz, 144.275–144.300 MHz, 222.05–222.06 MHz or 432.300–432.400 MHz segments, or on the 33 cm and shorter wavelength bands.

(e) Before establishing an automatically controlled beacon in the National Radio Quiet Zone or before changing the transmitting frequency, transmitter power, antenna height or directivity, the station licensee must give written notification thereof to the Interference Office, National Radio Astronomy Observatory, P.O. Box 2, Green Bank, WV 24944.

(1) The notification must include the geographical coordinates of the antenna, antenna ground elevation above mean sea level (AMSL), antenna center of radiation above ground level (AGL), antenna directivity, proposed frequency, type of emission, and transmitter power.

(2) If an objection to the proposed operation is received by the FCC from the National Radio Astronomy Observatory at Green Bank, Pocahontas County, WV, for itself or on

behalf of the Naval Research Laboratory at Sugar Grove, Pendleton County, WV, within 20 days from the date of notification, the FCC will consider all aspects of the problem and take whatever action is deemed appropriate.

(f) A beacon must cease transmissions upon notification by a District Director that the station is operating improperly or causing undue interference to other operations. The beacon may not resume transmitting without prior approval of the District Director.

(g) A beacon may transmit one-way communications.

§ 97.205 Repeater station.

(a) Any amateur station licensed to a holder of a Technician, General, Advanced or Amateur Extra Class operator license may be a repeater. A holder of a Technician, General, Advanced or Amateur Extra Class operator license may be the control operator of a repeater, subject to the privileges of the class of operator license held.

(b) A repeater may receive and retransmit only on the 10 m and shorter wavelength frequency bands except the 28.0–29.5 MHz, 50.0–51.0 MHz, 144.0–144.5 MHz, 145.5–146.0 MHz, 222.00–222.15 MHz, 431.0–433.0 Mhz, and 435.0–438.0 Mhz segments.

(c) Where the transmissions of a repeater cause harmful interference to another repeater, the two station licensees are equally and fully responsible for resolving the interference unless the operation of one station is recommended by a frequency coordinator and the operation of the other station is not. In that case, the licensee of the non-coordinated repeater has primary responsibility to resolve the interference.

(d) A repeater may be automatically controlled.

(e) Ancillary functions of a repeater that are available to users on the input channel are not considered remotely controlled functions of the station. Limiting the use of a repeater to only certain user stations is permissible.

(f) [Reserved]

(g) The control operator of a repeater that retransmits inadvertently communications that violate the rules in this part is not accountable for the violative communications.

(h) The provisions of this paragraph do not apply to repeaters that transmit on the 1.2 cm or shorter wavelength bands. Before establishing a repeater within 16 km (10 miles) of the Arecibo Observatory or before changing the transmitting frequency, transmitter power, antenna height or directivity of an existing repeater, the station licensee must give written notification thereof to the Interference Office, Arecibo Observatory, HC3 Box 53995, Arecibo, Puerto Rico 00612, in writing or electronically, of the technical parameters of the proposal. Licensees who choose to transmit information electronically should e-mail to: *prcz@naic.edu.*

(1) The notification shall state the geographical coordinates of the antenna (NAD–83 datum), antenna height above mean sea level (AMSL), antenna center of radiation

above ground level (AGL), antenna directivity and gain, proposed frequency and FCC Rule Part, type of emission, effective radiated power, and whether the proposed use is itinerant. Licensees may wish to consult interference guidelines provided by Cornell University.

(2) If an objection to the proposed operation is received by the FCC from the Arecibo Observatory, Arecibo, Puerto Rico, within 20 days from the date of notification, the FCC will consider all aspects of the problem and take whatever action is deemed appropriate. The licensee will be required to make reasonable efforts in order to resolve or mitigate any potential interference problem with the Arecibo Observatory.

§ 97.207 Space station.

(a) Any amateur station may be a space station. A holder of any class operator license may be the control operator of a space station, subject to the privileges of the class of operator license held by the control operator.

(b) A space station must be capable of effecting a cessation of transmissions by telecommand whenever such cessation is ordered by the FCC.

(c) The following frequency bands and segments are authorized to space stations:

(1) The 17 m, 15 m, 12 m, and 10 m bands, 6 mm, 4 mm, 2 mm and 1 mm bands; and

(2) The 7.0–7.1 MHz, 14.00–14.25 MHz, 144–146 MHz, 435–438 MHz, 2400–2450 MHz, 3.40–3.41 GHz, 5.83–5.85 GHz, 10.45–10.50 GHz, and 24.00–24.05 GHz segments.

(d) A space station may automatically retransmit the radio signals of Earth stations and other space stations.

(e) A space station may transmit one-way communications.

(f) Space telemetry transmissions may consist of specially coded messages intended to facilitate communications or related to the function of the spacecraft.

(g) The license grantee of each space station must make the following written notifications to the International Bureau, FCC, Washington, DC 20554.

(1) A pre-space notification within 30 days after the date of launch vehicle determination, but no later than 90 days before integration of the space station into the launch vehicle. The notification must be in accordance with the provisions of Articles 9 and 11 of the International Telecommunication Union (ITU) Radio Regulations and must specify the information required by Appendix 4 and Resolution No. 642 of the ITU Radio Regulations. The notification must also include a description of the design and operational strategies that the space station will use to mitigate orbital debris, including the following information:

(i) A statement that the space station licensee has assessed and limited the amount of debris released in a planned manner during normal operations, and has assessed and limited the probability of the space station becoming a source of debris by collisions with small debris or meteoroids that could cause loss of control and prevent post-mission disposal;

(ii) A statement that the space station licensee has assessed and limited the probability of accidental explosions during and after completion of mission operations. This statement must include a demonstration that debris generation will not result from the conversion of energy sources on board the spacecraft into energy that fragments the spacecraft. Energy sources include chemical, pressure, and kinetic energy. This demonstration should address whether stored energy will be removed at the spacecraft's end of life, by depleting residual fuel and leaving all fuel line valves open, venting any pressurized system, leaving all batteries in a permanent discharge state, and removing any remaining source of stored energy, or through other equivalent procedures specifically disclosed in the application;

(iii) A statement that the space station licensee has assessed and limited the probability of the space station becoming a source of debris by collisions with large debris or other operational space stations. Where a space station will be launched into a low-Earth orbit that is identical, or very similar, to an orbit used by other space stations, the statement must include an analysis of the potential risk of collision and a description of what measures the space station operator plans to take to avoid in-orbit collisions. If the space station licensee is relying on coordination with another system, the statement must indicate what steps have been taken to contact, and ascertain the likelihood of successful coordination of physical operations with, the other system. The statement must disclose the accuracy—if any—with which orbital parameters of non-geostationary satellite orbit space stations will be maintained, including apogee, perigee, inclination, and the right ascension of the ascending node(s). In the event that a system is not able to maintain orbital tolerances, *i.e.*, it lacks a propulsion system for orbital maintenance, that fact should be included in the debris mitigation disclosure. Such systems must also indicate the anticipated evolution over time of the orbit of the proposed satellite or satellites. Where a space station requests the assignment of a geostationary-Earth orbit location, it must assess whether there are any known satellites located at, or reasonably expected to be located at, the requested orbital location, or assigned in the vicinity of that location, such that the station keeping volumes of the respective satellites might overlap. If so, the statement must include a statement as to the identities of those parties and the measures that will be taken to prevent collisions;

(iv) A statement detailing the post-mission disposal plans for the space station at end of life, including the quantity of fuel—if any—that will be reserved for post-mission disposal maneuvers. For geostationary-Earth orbit space stations, the statement must disclose the altitude selected for a post-mission disposal orbit and the calculations that are used in deriving the disposal altitude. The statement must also include a casualty risk assessment if planned post-mission disposal involves atmospheric re-entry of the space station. In general, an assessment should include an estimate as to whether portions of the spacecraft will survive re-entry and reach the surface of the Earth, as well as an estimate of the resulting probability of human casualty.

(v) If any material item described in this notification changes before launch, a replacement pre-space notification shall be filed with the International Bureau no later than 90 days before integration of the space station into the launch vehicle.

(2) An in-space station notification is required no later than 7 days following initiation of space station transmissions. This notification must update the information contained in the pre-space notification.

(3) A post-space station notification is required no later than 3 months after termination of the space station transmissions. When termination of transmissions is ordered by the FCC, the notification is required no later than 24 hours after termination of transmissions.

§ 97.209 Earth station.

(a) Any amateur station may be an Earth station. A holder of any class operator license may be the control operator of an Earth station, subject to the privileges of the class of operator license held by the control operator.

(b) The following frequency bands and segments are authorized to Earth stations:

(1) The 17 m, 15 m, 12 m, and 10 m bands, 6 mm, 4 mm, 2 mm and 1 mm bands; and

(2) The 7.0–7.1 MHz, 14.00–14.25 MHz, 144–146 MHz, 435–438 MHz, 1260–1270 MHz and 2400–2450 MHz, 3.40–3.41 GHz, 5.65–

5.67 GHz, 10.45–10.50 GHz and 24.00–24.05 GHz segments.

§ 97.211 Space telecommand station.

(a) Any amateur station designated by the licensee of a space station is eligible to transmit as a telecommand station for that space station, subject to the privileges of the class of operator license held by the control operator.

(b) A telecommand station may transmit special codes intended to obscure the meaning of telecommand messages to the station in space operation.

(c) The following frequency bands and segments are authorized to telecommand stations:

(1) The 17 m, 15 m, 12 m and 10 m bands, 6 mm, 4 mm, 2 mm and 1 mm bands; and

(2) The 7.0–7.1 MHz, 14.00–14.25 MHz, 144–146 MHz, 435–438 MHz, 1260–1270 MHz and 2400–2450 MHz, 3.40–3.41 GHz, 5.65–

5.67 GHz, 10.45–10.50 GHz and 24.00–24.05 GHz segments.

(d) A telecommand station may transmit one-way communications.

§ 97.213 Telecommand of an amateur station.

An amateur station on or within 50 km of the Earth's surface may be under telecommand where:

(a) There is a radio or wireline control link between the control point and the station sufficient for the control operator to perform his/her duties. If radio, the control link must use an auxiliary station. A control link using a fiber optic cable or another telecommunication service is considered wireline.

(b) Provisions are incorporated to limit transmission by the station to a period of no more than 3 minutes in the event of malfunction in the control link.

(c) The station is protected against making, willfully or negligently, unauthorized transmissions.

(d) A photocopy of the station license and a label with the name, address, and telephone number of the station licensee and at least one designated control operator is posted in a conspicuous place at the station location.

§ 97.215 Telecommand of model craft.

An amateur station transmitting signals to control a model craft may be operated as follows:

(a) The station identification procedure is not required for transmissions directed only to the model craft, provided that a label indicating the station call sign and the station licensee's name and address is affixed to the station transmitter.

(b) The control signals are not considered codes or ciphers intended to obscure the meaning of the communication.

(c) The transmitter power must not exceed 1 W.

§ 97.217 Telemetry.

Telemetry transmitted by an amateur station on or within 50 km of the Earth's surface is not considered to be codes or ciphers intended to obscure the meaning of communications.

§ 97.219 Message forwarding system.

(a) Any amateur station may participate in a message forwarding system, subject to the privileges of the class of operator license held.

(b) For stations participating in a message forwarding system, the control operator of the station originating a message is primarily accountable for any violation of the rules in this part contained in the message.

(c) Except as noted in (d) of this section, for stations participating in a message forwarding system, the control operators of forwarding stations that retransmit inadvertently communications that violate the rules in this part are not accountable for the violative communications. They are, however, responsible for discontinuing such communications once they become aware of their presence.

(d) For stations participating in a message forwarding system, the control operator of the first forwarding station must:

(1) Authenticate the identity of the station from which it accepts communications on

behalf of the system; or

(2) Accept accountability for any violation of the rules in this part contained in messages it retransmits to the system.

§ 97.221 Automatically controlled digital station.

(a) This rule section does not apply to an auxiliary station, a beacon station, a repeater station, an earth station, a space station, or a space telecommand station.

(b) A station may be automatically controlled while transmitting a RTTY or data emission on the 6 m or shorter wavelength bands, and on the 28.120–28.189 MHz, 24.925–24.930 MHz, 21.090–21.100 MHz, 18.105–18.110 MHz, 14.0950–14.0995 MHz, 14.1005–14.112 MHz, 10.140–10.150 MHz, 7.100–7.105 MHz, or 3.585–3.600 MHz segments.

(c) A station may be automatically controlled while transmitting a RTTY or data emission on any other frequency authorized for such emission types provided that:

(1) The station is responding to interrogation by a station under local or remote control; and

(2) No transmission from the automatically controlled station occupies a bandwidth of more than 500 Hz.

Subpart D—Technical Standards § 97.301 Authorized frequency bands.

The following transmitting frequency bands are available to an amateur station located within 50 km of the Earth's surface, within the specified ITU Region, and outside any area where the amateur service is regulated by any authority other than the FCC.

(a) For a station having a control operator who has been granted a Technician, General, Advanced, or Amateur Extra Class operator license or who holds a CEPT radio-amateur license or IARP of any class:

(b) For a station having a control operator who has been granted an Amateur Extra Class operator license, who holds a CEPT radio amateur license, or who holds a Class 1 IARP license:

(c) For a station having a control operator who has been granted an operator license of Advanced Class:

Wavelength band	Evaluation required if power1(watts) exceeds
MF	

Band	Evaluation threshold (W)
160 m	500
HF	
80 m	500
75 m	500
40 m	500
30 m	425
20 m	225
17 m	125
15 m	100
12 m	75
10 m	50
VHF (all bands)	50

Band	Evaluation threshold (W)
UHF	
70 cm	70
33 cm	150
23 cm	200
13 cm	250
SHF (all bands)	250
EHF (all bands)	250
Repeater stations (all bands)	non-building-mounted antennas: height above ground level to lowest point of antenna <10 m and power >500 W ERP building-mounted antennas: power >500 W ERP

(d) For a station having a control operator who has been granted an operator license of General Class:

(e) For a station having a control operator who has been granted an operator license of Novice Class or Technician Class:

The following paragraphs summarize the frequency sharing requirements that apply to amateur stations transmitting in the frequency bands specified in §97.301 of this part. Each frequency band allocated to the amateur service is designated as either a secondary service or a primary service. A station in a secondary service must not cause harmful interference to, and must accept interference from, stations in a primary service.

(a) Where, in adjacent ITU Regions or sub-Regions, a band of frequencies is allocated to different services of the same category (*i.e.*, primary or secondary services), the basic principle is the equality of right to operate. Accordingly, stations of each service in one Region or sub-Region must operate so as not to cause harmful interference to any service of the same or higher category in the other Regions or sub-Regions.

(b) Amateur stations transmitting in the 70 cm band, the 33 cm band, the 23 cm band, the 9 cm band, the 5 cm band, the 3 cm band, or the 24.05–24.25 GHz segment must not cause harmful interference to, and must accept interference from, stations authorized by the United States Government in the radiolocation service.

(c) Amateur stations transmitting in the 1900–2000 kHz segment, the 76–77.5 GHz segment, the 78–81 GHz segment, the 136–141 GHz segment, or the 241–248 GHz segment must not cause harmful interference to, and must accept interference from, stations authorized by the United States Government, the FCC, or other nations in the radiolocation service.

(d) Amateur stations transmitting in the 430–450 MHz segment, the 23 cm band, the 3.3–3.4 GHz segment, the 5.65–5.85 GHz segment, the 13 cm band, or the 24.05–24.25 GHz segment, must not cause harmful interference to, and must accept interference from, stations authorized by other nations in the radiolocation service.

(e) Amateur stations receiving in the 33 cm band, the 2400–2450 MHz segment, the 5.725–5.875 GHz segment, the 1.2 cm band, the

Wavelength band	ITU region 1	ITU region 2	ITU region 3	Sharing requirements *see* §97.303 (paragraph)
MF	kHz	kHz	kHz	

The Complete Study Guide For All Amateur Radio Tests

Wavelength	ITU	ITU	ITU	Sharing requirements see
160 m	1810–1850	1800–2000	1800–2000	(a), (c), (g)
HF	**MHz**	**MHz**	**MHz**	
80 m	3.525–3.600	3.525–3.600	3.525–3.600	(a)
75 m		3.800–4.000	3.800–3.900	(a)
60 m		See §97.303(h)		(h)
40 m	7.025–7.125	7.025–7.125	7.025–7.125	(i)
Do	7.175–7.200	7.175–7.300	7.175–7.200	(i)
30 m	10.100–10.150	10.100–10.150	10.100–10.150	(j)
20 m	14.025–14.150	14.025–14.150	14.025–14.150	
Do	14.225–14.350	14.225–14.350	14.225–14.350	
17 m	18.068–18.168	18.068–18.168	18.068–18.168	
15 m	21.025–21.200	21.025–21.200	21.025–21.200	
Do	21.275–21.450	21.275–21.450	21.275–21.450	
12 m	24.890–24.990	24.890–24.990	24.890–24.990	
10 m	28.000–29.700	28.000–29.700	28.000–29.700	

band	region 1	region 2	region 3	§97.303 (paragraph)
HF	MHz	MHz	MHz	
80 m	3.525–3.600	3.525–3.600	3.525–3.600	(a)
40 m	7.025–7.125	7.025–7.125	7.025–7.125	(i)
15 m	21.025–21.200	21.025–21.200	21.025–21.200	
10 m	28.0–28.5	28.0–28.5	28.0–28.5	
VHF	MHz	MHz	MHz	
1.25 m		222–225		(a)
UHF	MHz	MHz	MHz	
23 cm	1270–1295	1270–1295	1270–1295	(d), (o)

	Frequencies	Emission types authorized	Standards see §97.307(f), paragraph:
MF:			
160 m	Entire band	RTTY, data	(3).
160 m	Entire band	Phone, image	(1), (2).
HF:			
80 m	Entire band	RTTY, data	(3), (9).
75 m	Entire band	Phone, image	(1), (2).
40 m	7.000–7.100 MHz	RTTY,	(3), (9)

		data	
40 m	7.075–7.100 MHz	Phone, image	(1), (2), (9), (11)

40 m	7.100–7.125 MHz	RTTY, data	(3), (9)
40 m	7.125–7.300 MHz	Phone, image	(1), (2)
30 m	Entire band	RTTY, data	(3).
20 m	14.00–14.15 MHz	RTTY, data	(3).
20 m	14.15–14.35 MHz	Phone, image	(1), (2).
17 m	18.068–18.110 MHz	RTTY, data	(3).
17 m	18.110–18.168 MHz	Phone, image	(1), (2).
15 m	21.0–21.2 MHz	RTTY, data	(3), (9).
15 m	21.20–21.45 MHz	Phone, image	(1), (2).
12 m	24.89–24.93 MHz	RTTY, data	(3).
12 m	24.93–24.99 MHz	Phone, image	(1), (2).
10 m	28.0–28.3 MHz	RTTY, data	(4).
10 m	28.3–28.5 MHz	Phone, image	(1), (2), (10).
10	28.5–29.0	Phone,	(1), (2).

	m	MHz	image	
	10 m	29.0–29.7 MHz	Phone, image	(2).
VHF:				
	6 m	50.1–51.0 MHz	MCW, phone, image, RTTY, data	(2), (5).
	Do	51.0–54.0 MHz	MCW, phone, image, RTTY, data, test	(2), (5), (8).
	2 m	144.1–148.0 MHz	MCW, phone, image, RTTY, data, test	(2), (5), (8).
1.25 m		219–220 MHz	Data	(13)
	Do	222–225 MHz	RTTY, data, test MCW, phone, SS, image	(2), (6), (8)
UHF:				
	70 cm	Entire band	MCW, phone, image, RTTY, data, SS, test	(6), (8).
	33 cm	Entire band	MCW, phone, image, RTTY, data, SS, test, pulse	(7), (8), and (12).

	23 cm	Entire band	MCW, phone, image, RTTY, data, SS, test	(7), (8), and (12).
	13 cm	Entire band	MCW, phone, image, RTTY, data, SS, test, pulse	(7), (8), and (12).
SHF:				
	9 cm	Entire band	MCW, phone, image, RTTY, data, SS, test, pulse	(7), (8), and (12).

			test, pulse	
	5 cm	Entire band	MCW, phone, image, RTTY, data, SS, test, pulse	(7), (8), and (12).
	3 cm	Entire band	MCW, phone, image, RTTY, data, SS, test	(7), (8), and (12).
	1.2 cm	Entire band	MCW, phone, image, RTTY, data, SS, test, pulse	(7), (8), and (12).
EHF:				
	6 mm	Entire band	MCW, phone, image, RTTY, data, SS, test, pulse	(7), (8), and (12).
	4 mm	Entire band	MCW, phone, image, RTTY, data, SS, test, pulse	(7), (8), and (12).
	2.5 mm	Entire band	MCW, phone, image, RTTY, data, SS, test, pulse	(7), (8), and (12).
	2	Entire band	MCW, phone, image, RTTY,	(7), (8),

	mm		data, SS, test, pulse	and (12).
	1mm	Entire band	MCW, phone, image, RTTY, data, SS, test, pulse	(7), (8), and (12).
		Above 275 GHz	MCW, phone, image, RTTY, data, SS, test, pulse	(7), (8), and (12).

2.5 mm band, or the 244–246 GHz segment must accept interference from industrial, scientific, and medical (ISM) equipment. (f) Amateur stations transmitting in the following segments must not cause harmful interference to radio astronomy stations: 3.332– 3.339 GHz, 3.3458–3.3525 GHz, 76–77.5 GHz, 78–81 GHz, 136–141 GHz, 241–248 GHz, 275–323 GHz, 327–371 GHz, 388–424 GHz, 426–442 GHz, 453–510 GHz, 623–711 GHz, 795–909 GHz, or 926–945 GHz. In addition, amateur stations transmitting in the following segments must not cause harmful interference to stations in the Earth exploration-satellite service (passive) or the space research service (passive): 275–277 GHz, 294–306 GHz, 316–334 GHz, 342–349 GHz, 363–365 GHz, 371–389 GHz, 416–434 GHz, 442–444 GHz, 496–506 GHz, 546–568 GHz, 624–629 GHz, 634–654 GHz, 659–661 GHz, 684–692 GHz, 730–732 GHz, 851–853 GHz, or 951–956 GHz.

(g) Amateur stations transmitting in the 1900–2000 kHz segment must not cause harmful interference to, and must accept interference from, stations authorized by other nations in the fixed, mobile except aeronautical mobile, and radionavigation services.

(h) Amateur stations may only transmit single sideband, suppressed carrier (emission type 2K80J3E), upper sideband on the channels 5332 kHz, 5348 kHz, 5368 kHz, 5373 kHz, and 5405 kHz. Amateur operators shall ensure that their station's transmission occupies only 2.8 kHz centered at each of these frequencies. Amateur stations must not cause harmful interference to, and must accept interference from, stations authorized by:

(1) The United States Government, the FCC, or other nations in the fixed service; and

(2) Other nations in the mobile except aeronautical mobile service.

(i) Amateur stations transmitting in the 7.2–7.3 MHz segment must not cause harmful interference to, and must accept interference from, international broadcast stations whose programming is intended for use within Region 1 or Region 3.

(j) Amateur stations transmitting in the 30 m band must not cause harmful interference to, and must accept interference from, stations by other nations in the fixed service. The licensee of the amateur station must make all necessary adjustments, including termination of transmissions, if harmful interference is caused.

(k) For amateur stations located in ITU Regions 1 and 3: Amateur stations transmitting in the 146–148 MHz segment or the 10.00–

10.45 GHz segment must not cause harmful interference to, and must accept interference from, stations of other nations in the fixed and mobile services.

(l) *In the 219–220 MHz segment:*

(1) Use is restricted to amateur stations participating as forwarding stations in fixed point-to-point digital message forwarding systems, including intercity packet backbone networks. It is not available for other purposes.

(2) Amateur stations must not cause harmful interference to, and must accept interference from, stations authorized by:

(i) The FCC in the Automated Maritime Telecommunications System (AMTS), the 218–219 MHz Service, and the 220 MHz Service, and television stations broadcasting on channels 11 and 13; and

(ii) Other nations in the fixed and maritime mobile services.

(3) No amateur station may transmit unless the licensee has given written notification of the station's specific geographic location for such transmissions in order to be incorporated into a database that has been made available to the public. The notification must be given at least 30 days prior to making such transmissions. The notification must be given to: The American Radio Relay League, Inc., 225 Main Street, Newington, CT 06111–1494.

(4) No amateur station may transmit from a location that is within 640 km of an AMTS coast station that operates in the 217–218 MHz and 219–220 MHz bands unless the amateur station licensee has given written notification of the station's specific geographic location for such transmissions to the AMTS licensee. The notification must be given at least 30 days prior to making such transmissions. The location of AMTS coast stations using the 217–218/219–220 MHz channels may be obtained as noted in paragraph (l)(3) of this section.

(5) No amateur station may transmit from a location that is within 80 km of an AMTS coast station that uses frequencies in the 217–218 MHz and 219–220 MHz bands unless that amateur station licensee holds written approval from that AMTS licensee. The location of AMTS coast stations using the 217–218/219–220 MHz channels may be obtained as noted in paragraph (l)(3) of this section.

(m) *In the 70 cm band:*

(1) No amateur station shall transmit from north of Line A in the 420–430 MHz segment. See §**97**.3(a) for the definition of Line A.

(2) Amateur stations transmitting in the 420–430 MHz segment must not cause harmful interference to, and must accept interference from, stations authorized by the FCC in the land mobile service within 80.5 km of Buffalo, Cleveland, and Detroit. See §2.106,

footnote US230 for specific frequencies and coordinates.

(3) Amateur stations transmitting in the 420–430 MHz segment or the 440–450 MHz segment must not cause harmful interference to, and must accept interference from, stations authorized by other nations in the fixed and mobile except aeronautical mobile services.

(n) *In the 33 cm band:*

(1) Amateur stations must not cause harmful interference to, and must accept interference from, stations authorized by:

(i) The United States Government;

(ii) The FCC in the Location and Monitoring Service; and

(iii) Other nations in the fixed service.

(2) No amateur station shall transmit from those portions of Texas and New Mexico that are bounded by latitudes 31°41' and 34°30' North and longitudes 104°11' and 107°30' West; or from outside of the United States and its Region 2 insular areas.

(3) No amateur station shall transmit from those portions of Colorado and Wyoming that are bounded by latitudes 39° and 42° North and longitudes 103° and 108° West in the following segments: 902.4–902.6 MHz, 904.3–904.7 MHz, 925.3–925.7 MHz, and 927.3–927.7 MHz.

(o) Amateur stations transmitting in the 23 cm band must not cause harmful interference to, and must accept interference from, stations authorized by:

(1) The United States Government in the aeronautical radionavigation, Earth exploration-satellite (active), or space research (active) services;

(2) The FCC in the aeronautical radionavigation service; and

(3) Other nations in the Earth exploration-satellite (active), radionavigation-satellite (space-to-Earth) (space-to-space), or space research (active) services.

(p) *In the 13 cm band:*

(1) Amateur stations must not cause harmful interference to, and must accept interference from, stations authorized by other nations in fixed and mobile services.

(2) Amateur stations transmitting in the 2305–2310 MHz segment must not cause harmful interference to, and must accept interference from, stations authorized by the FCC in the fixed, mobile except aeronautical mobile, and radiolocation services.

(q) Amateur stations transmitting in the 3.4–3.5 GHz segment must not cause harmful interference to, and must accept interference from, stations authorized by other nations in the fixed and fixed-satellite (space-to-Earth) services.

(r) *In the 5 cm band:*

(1) Amateur stations transmitting in the 5.650–5.725 GHz segment must not cause harmful interference to, and must accept interference from, stations authorized by other nations in the mobile except aeronautical mobile service.

(2) Amateur stations transmitting in the 5.850–5.925 GHz segment must not cause harmful interference to, and must accept interference from, stations authorized by the FCC and other nations in the fixed-satellite (Earth-to-space) and mobile services and also stations authorized by other nations in the fixed service. In the United States, the use of mobile service is restricted to Dedicated Short Range Communications operating in the Intelligent Transportation System.

(s) Authorization of the 76–77 GHz segment for amateur station transmissions is suspended until such time that the Commission may determine that amateur station transmissions in this segment will not pose a safety threat to vehicle radar systems operating in this segment.

(t) Amateur stations transmitting in the 2.5 mm band must not cause harmful interference to, and must accept interference from, stations authorized by the United States Government, the FCC, or other nations in the fixed, inter-satellite, or mobile services.

Note to §97.303: The Table of Frequency Allocations contains the complete, unabridged, and legally binding frequency sharing requirements that pertain to the Amateur Radio Service. *See* 47 CFR 2.104, 2.105, and 2.106. The United States, Puerto Rico, and the U.S. Virgin Islands are in Region 2 and other U.S. insular areas are in either Region 2 or 3; see appendix 1 to part **97**.

§ 97.305 Authorized emission types.

(a) Except as specified elsewhere in this part, an amateur station may transmit a CW emission on any frequency authorized to the control operator.

(b) A station may transmit a test emission on any frequency authorized to the control operator for brief periods for experimental purposes, except that no pulse modulation emission may be transmitted on any frequency where pulse is not specifically authorized and no SS modulation emission may be transmitted on any frequency where SS is not specifically authorized.

(c) A station may transmit the following emission types on the frequencies indicated, as authorized to the control operator, subject to the standards specified in §97.307(f) of this part.

(a) No amateur station transmission shall occupy more bandwidth than necessary for the information rate and emission type being transmitted, in accordance with good amateur practice.

(b) Emissions resulting from modulation must be confined to the band or segment available to the control operator. Emissions outside the necessary bandwidth must not cause splatter or keyclick interference to operations on adjacent frequencies.

(c) All spurious emissions from a station transmitter must be reduced to the greatest extent practicable. If any spurious emission, including chassis or power line radiation, causes harmful interference to the reception of another radio station, the licensee of the interfering amateur station is required to take steps to eliminate the interference, in accordance with good engineering practice.

(d) For transmitters installed after January 1, 2003, the mean power of any spurious emission from a station transmitter or external RF power amplifier transmitting on a frequency below 30 MHz must be at least 43 dB below the mean power of the fundamental emission. For transmitters installed on or before January 1, 2003, the mean power of any spurious emission from a station transmitter or external RF power amplifier transmitting on a frequency below 30 MHz must not exceed 50 mW and must be at least 40 dB below the mean power of the fundamental emission. For a transmitter of mean power less than 5 W installed on or before January 1, 2003, the attenuation must be at least 30 dB. A transmitter built before April 15, 1977, or first marketed before January 1, 1978, is exempt from this requirement.

(e) The mean power of any spurious emission from a station transmitter or external RF power amplifier transmitting on a frequency between 30–225 MHz must be at least 60 dB below the mean power of the fundamental. For a transmitter having a mean power of 25 W or less, the mean power of any spurious emission supplied to the antenna transmission line must not exceed 25 μW and must be at least 40 dB below the mean power of the fundamental emission, but need not be reduced below the power of 10 μW. A transmitter built before April 15, 1977, or first marketed before January 1, 1978, is exempt from this requirement.

(f) The following standards and limitations apply to transmissions on the frequencies specified in §**97**.305(c) of this part.

(1) No angle-modulated emission may have a modulation index greater than 1 at the highest modulation frequency.

(2) No non-phone emission shall exceed the bandwidth of a communications quality phone emission of the same modulation type. The total bandwidth of an independent sideband emission (having B as the first symbol), or a multiplexed image and phone emission, shall not exceed that of a communications quality A3E emission.

(3) Only a RTTY or data emission using a specified digital code listed in §**97**.309(a) of this part may be transmitted. The symbol rate must not exceed 300 bauds, or for frequency-shift keying, the frequency shift between mark and space must not exceed 1 kHz.

(4) Only a RTTY or data emission using a specified digital code listed in §**97**.309(a) of this part may be transmitted. The symbol rate must not exceed 1200 bauds, or for frequency-shift keying, the frequency shift between mark and space must not exceed 1 kHz.

(5) A RTTY, data or multiplexed emission using a specified digital code listed in §**97**.309(a) of this part may be transmitted. The symbol rate must not exceed 19.6 kilobauds. A RTTY, data or multiplexed emission using an unspecified digital code under the limitations listed in §**97**.309(b) of this part also may be transmitted. The

authorized bandwidth is 20 kHz.

(6) A RTTY, data or multiplexed emission using a specified digital code listed in §97.309(a) of this part may be transmitted. The symbol rate must not exceed 56 kilobauds. A RTTY, data or multiplexed emission using an unspecified digital code under the limitations listed in §97.309(b) of this part also may be transmitted. The authorized bandwidth is 100 kHz.

(7) A RTTY, data or multiplexed emission using a specified digital code listed in §97.309(a) of this part or an unspecified digital code under the limitations listed in §97.309(b) of this part may be transmitted.

(8) A RTTY or data emission having designators with A, B, C, D, E, F, G, H, J or R as the first symbol; 1, 2, 7 or 9 as the second symbol; and D or W as the third symbol is also authorized.

(9) A station having a control operator holding a Novice or Technician Class operator license may only transmit a CW emission using the international Morse code.

(10) A station having a control operator holding a Novice Class operator license or a Technician Class operator license and who has received credit for proficiency in telegraphy in accordance with the international requirements may only transmit a CW emission using the international Morse code or phone emissions J3E and R3E.

(11) Phone and image emissions may be transmitted only by stations located in ITU Regions 1 and 3, and by stations located within ITU Region 2 that are west of 130° West longitude or south of 20° North latitude.

(12) Emission F8E may be transmitted.

(13) A data emission using an unspecified digital code under the limitations listed in §97.309(b) also may be transmitted. The authorized bandwidth is 100 kHz.

Wavelength band

---

§ 97.307 Emission standards.

§ 97.309 RTTY and data emission codes.

(a) Where authorized by §§97.305(c) and 97.307(f) of the part, an amateur station may transmit a RTTY or data emission using the following specified digital codes:

(1) The 5-unit, start-stop, International Telegraph Alphabet No. 2, code defined in ITU–T Recommendation F.1, Division C (commonly known as "Baudot").

(2) The 7-unit code specified in ITU–R Recommendations M.476–5 and M.625–3 (commonly known as "AMTOR").

(3) The 7-unit, International Alphabet No. 5, code defined in IT–T Recommendation T.50 (commonly known as "ASCII").

(4) An amateur station transmitting a RTTY or data emission using a digital code specified in this paragraph may use any technique whose technical characteristics have been documented publicly, such as CLOVER, G-TOR, or PacTOR, for the purpose of facilitating communications.

(b) Where authorized by §§97.305(c) and 97.307(f) of this part, a station may transmit a RTTY or data emission using an unspecified digital code, except to a station in a country with which the United States does not have an agreement permitting the code to be used. RTTY and data emissions using unspecified digital codes must not be transmitted for the purpose of obscuring the meaning of any communication. When deemed necessary by a District Director to assure compliance with the FCC Rules, a station must:

(1) Cease the transmission using the unspecified digital code;

(2) Restrict transmissions of any digital code to the extent instructed;

(3) Maintain a record, convertible to the original information, of all digital communications transmitted.

§ 97.311 SS emission types.

(a) SS emission transmissions by an amateur station are authorized only for communications between points within areas where the amateur service is regulated by the FCC and between an area where the amateur service is regulated by the FCC and an amateur station in another country that permits such communications. SS emission transmissions must not be used for the purpose of obscuring the meaning of any communication.

(b) A station transmitting SS emissions must not cause harmful interference to stations employing other authorized emissions, and must accept all interference caused by stations employing other authorized emissions.

(c) When deemed necessary by a District Director to assure compliance with this part, a station licensee must:

(1) Cease SS emission transmissions;

(2) Restrict SS emission transmissions to the extent instructed; and

(3) Maintain a record, convertible to the original information (voice, text, image, etc.) of all spread spectrum communications transmitted.

§ 97.313 Transmitter power standards.

(a) An amateur station must use the minimum transmitter power necessary to carry out the desired communications.

(b) No station may transmit with a transmitter power exceeding 1.5 kW PEP.

(c) No station may transmit with a transmitter power output exceeding 200 W PEP:

(1) On the 10.10–10.15 MHz segment;

(2) On the 3.525–3.60 MHz, 7.025–7.125 MHz, 21.025–21.20 MHz, and 28.0–28.5 MHz segment when the control operator is a Novice Class operator or a Technician Class operator; or

(3) The 7.050-7.075 MHz segment when the station is within ITU Regions 1 or 3.

(d) No station may transmit with a transmitter power exceeding 25 W PEP on the VHF 1.25 m band when the control operator is a Novice operator.

(e) No station may transmit with a transmitter power exceeding 5 W PEP on the UHF 23 cm band when the control operator is a Novice operator.

(f) No station may transmit with a transmitter power exceeding 50 W PEP on the UHF 70 cm band from an area specified in footnote US7 to §2.106 of part 2, unless expressly authorized by the FCC after mutual agreement, on a case-by-case basis, between the District Director of the applicable field facility and the military area frequency coordinator at the applicable military base. An Earth station or telecommand station, however, may transmit on the 435–438 MHz segment with a maximum of 611 W effective radiated power (1 kW equivalent isotropically radiated power) without the authorization otherwise required. The transmitting antenna elevation angle between the lower half-power (−3 dB relative to the peak or antenna bore sight) point and the horizon must always be greater than 10°.

(g) No station may transmit with a transmitter power exceeding 50 W PEP on the 33 cm band from within 241 km of the boundaries of the White Sands Missile Range. Its boundaries are those portions of Texas and New Mexico bounded on the south by latitude 31°41' North, on the east by longitude 104°11' West, on the north by latitude 34°30' North, and on the west by longitude 107°30' West.

(h) No station may transmit with a transmitter power exceeding 50 W PEP on the 219–220 MHz segment of the 1.25 m band.

(i) No station may transmit with an effective radiated power (ERP) exceeding 50 W PEP on the 60 m band. For the purpose of computing ERP, the transmitter PEP will be multiplied by the antenna gain relative to a dipole or the equivalent calculation in decibels.

A half-wave dipole antenna will be presumed to have a gain of 1. Licensees using other antennas must maintain in their station records either the antenna manufacturer data on the antenna gain or calculations of the antenna gain.

(j) No station may transmit with a transmitter output exceeding 10 W PEP when the station is transmitting a SS emission type.

§ 97.315 Certification of external RF power amplifiers.

(a) Any external RF power amplifier (see §2.815 of the FCC Rules) manufactured or imported for use at an amateur radio station must be certificated for use in the amateur service in accordance with subpart J of part 2 of the FCC Rules. No amplifier capable of operation below 144 MHz may be constructed or modified by a non-amateur service licensee without a grant of certification from the FCC.

(b) The requirement of paragraph (a) does not apply if one or more of the following conditions are met:

(1) The amplifier is constructed or modified by an amateur radio operator for use at an amateur station.

(2) The amplifier was manufactured before April 28, 1978, and has been issued a marketing waiver by the FCC, or the amplifier was purchased before April 28, 1978, by an amateur radio operator for use at that operator's station.

(3) The amplifier is sold to an amateur radio operator or to a dealer, the amplifier is purchased in used condition by a dealer, or the amplifier is sold to an amateur radio operator for use at that operator's station.

(c) Any external RF power amplifier appearing in the Commission's database as certificated for use in the amateur service may be marketed for use in the amateur service.

§ 97.317 Standards for certification of external RF power amplifiers.

(a) To receive a grant of certification, the amplifier must:

(1) Satisfy the spurious emission standards of §97.307 (d) or (e) of this part, as applicable, when the amplifier is operated at the lesser of 1.5 kW PEP or its full output power and when the amplifier is placed in the "standby" or "off" positions while connected to the transmitter.

(2) Not be capable of amplifying the input RF power (driving signal) by more than 15 dB gain. Gain is defined as the ratio of the input RF power to the output RF power of the amplifier where both power measurements are expressed in peak envelope power or mean power.

(3) Exhibit no amplification (0 dB gain) between 26 MHz and 28 MHz.

(b) Certification shall be denied when:

(1) The Commission determines the amplifier can be used in services other than the Amateur Radio Service, or

(2) The amplifier can be easily modified to operate on frequencies between 26 MHz and 28 MHz.

Subpart E—Providing Emergency Communications § 97.401 Operation during a disaster.

A station in, or within 92.6 km (50 nautical miles) of, Alaska may transmit emissions J3E and R3E on the channel at 5.1675 MHz (assigned frequency 5.1689 MHz) for emergency communications. The channel must be shared with stations licensed in the Alaska-Private Fixed Service. The transmitter power must not exceed 150 W PEP. A station in, or within 92.6 km of, Alaska may transmit communications for tests and training drills necessary to ensure the establishment, operation, and maintenance of emergency communication systems.

§ 97.403 Safety of life and protection of property.

No provision of these rules prevents the use by an amateur station of any means of radiocommunication at its disposal to provide essential communication needs in connection with the immediate safety of human life and immediate protection of property when normal communication systems are not available.

§ 97.405 Station in distress.

(a) No provision of these rules prevents the use by an amateur station in distress of any means at its disposal to attract attention, make known its condition and location, and obtain assistance.

(b) No provision of these rules prevents the use by a station, in the exceptional circumstances described in paragraph (a) of this section, of any means of radiocommunications at its disposal to assist a station in distress.

§ 97.407 Radio amateur civil emergency service.

(a) No station may transmit in RACES unless it is an FCC-licensed primary, club, or military recreation station and it is certified by a civil defense organization as registered with that organization. No person may be the control operator of an amateur station transmitting in RACES unless that person holds a FCC-issued amateur operator license and is certified by a civil defense organization as enrolled in that organization.

(b) The frequency bands and segments and emissions authorized to the control operator are available to stations transmitting communications in RACES on a shared basis with the amateur service. In the event of an emergency which necessitates invoking the President's War Emergency Powers under the provisions of section 706 of the Communications Act of 1934, as amended, 47 U.S.C. 606, amateur stations participating in RACES may only transmit on the frequency segments authorized pursuant to part 214 of this chapter.

(c) An amateur station registered with a civil defense organization may only communicate with the following stations upon authorization of the responsible civil defense official for the organization with which the amateur station is registered:

(1) An amateur station registered with the same or another civil defense organization; and

(2) A station in a service regulated by the FCC whenever such communication is authorized by the FCC.

(d) All communications transmitted in RACES must be specifically authorized by the civil defense organization for the area served. Only civil defense communications of the following types may be transmitted:

(1) Messages concerning impending or actual conditions jeopardizing the public safety, or affecting the national defense or security during periods of local, regional, or national civil emergencies;

(2) Messages directly concerning the immediate safety of life of individuals, the

immediate protection of property, maintenance of law and order, alleviation of human suffering and need, and the combating of armed attack or sabotage;

(3) Messages directly concerning the accumulation and dissemination of public information or instructions to the civilian population essential to the activities of the civil defense organization or other authorized governmental or relief agencies; and

(4) Communications for RACES training drills and tests necessary to ensure the establishment and maintenance of orderly and efficient operation of the RACES as ordered by the responsible civil defense organization served. Such drills and tests may not exceed a total time of 1 hour per week. With the approval of the chief officer for emergency planning in the applicable State, Commonwealth, District or territory, however, such tests and drills may be conducted for a period not to exceed 72 hours no more than twice in any calendar year.

Subpart F—Qualifying Examination Systems

§ 97.501 Qualifying for an amateur operator license.

Each applicant must pass an examination for a new amateur operator license grant and for each change in operator class. Each applicant for the class of operator license grant specified below must pass, or otherwise receive examination credit for, the following examination elements:

(a) Amateur Extra Class operator: Elements 2, 3, and 4;

(b) General Class operator: Elements 2 and 3;

(c) Technician Class operator: Element 2.

§ 97.503 Element standards.

A written examination must be such as to prove that the examinee possesses the operational and technical qualifications required to perform properly the duties of an amateur service licensee. Each written examination must be comprised of a question set as follows:

(a) Element 2: 35 questions concerning the privileges of a Technician Class operator license. The minimum passing score is 26 questions answered correctly.

(b) Element 3: 35 questions concerning the privileges of a General Class operator license. The minimum passing score is 26 questions answered correctly.

(c) Element 4: 50 questions concerning the privileges of an Amateur Extra Class operator license. The minimum passing score is 37 questions answered correctly.

§ 97.505 Element credit.

(a) The administering VEs must give credit as specified below to an examinee holding any of the following license grants or license documents:

(1) An unexpired (or expired but within the grace period for renewal) FCC-granted Advanced Class operator license grant: Elements 2 and 3.

(2) An unexpired (or expired but within the grace period for renewal) FCC-granted General Class operator license grant: Elements 2 and 3.

(3) An unexpired (or expired but within the grace period for renewal) FCC-granted Technician or Technician Plus Class operator (including a Technician Class operator license granted before February 14, 1991) license grant: Element 2.

(4) An expired FCC-issued Technician Class operator license document granted before March 21, 1987; Element 3.

(5) A CSCE: Each element the CSCE indicates the examinee passed within the previous 365 days.

(b) No examination credit, except as herein provided, shall be allowed on the basis of holding or having held any other license grant or document.

§ 97.507 Preparing an examination.

(a) Each telegraphy message and each written question set administered to an examinee must be prepared by a VE holding an Amateur Extra Class operator license. A telegraphy message or written question set may also be prepared for the following elements by a VE holding an operator license of the class indicated:

(1) Element 3: Advanced Class operator.

(2) Elements 1 and 2: Advanced or General Class operators.

(b) Each question set administered to an examinee must utilize questions taken from the applicable question pool.

(c) Each telegraphy message and each written question set administered to an examinee for an amateur operator license must be prepared, or obtained from a supplier, by the administering VEs according to instructions from the coordinating VEC.

(d) A telegraphy examination must consist of a message sent in the international Morse code at no less than the prescribed speed for a minimum of 5 minutes. The message must contain each required telegraphy character at least once. No message known to the examinee may be administered in a telegraphy examination. Each 5 letters of the alphabet must be counted as 1 word. Each numeral, punctuation mark and prosign must be counted as 2 letters of the alphabet.

§ 97.509 Administering VE requirements.

(a) Each examination for an amateur operator license must be administered by a team of at least 3 VEs at an examination session coordinated by a VEC. The number of examinees at the session may be limited.

(b) Each administering VE must:

(1) Be accredited by the coordinating VEC;

(2) Be at least 18 years of age;

(3) Be a person who holds an amateur operator license of the class specified below:

(i) Amateur Extra, Advanced or General Class in order to administer a Technician Class operator license examination;

(ii) Amateur Extra or Advanced Class in order to administer a General Class operator license examination;

(iii) Amateur Extra Class in order to administer an Amateur Extra Class operator license examination.

(4) Not be a person whose grant of an amateur station license or amateur operator license has ever been revoked or suspended.

(c) Each administering VE must be present and observing the examinee throughout the entire examination. The administering VEs are responsible for the proper conduct and necessary supervision of each examination. The administering VEs must immediately terminate the examination upon failure of the examinee to comply with their instructions.

(d) No VE may administer an examination to his or her spouse, children, grandchildren, stepchildren, parents, grandparents, stepparents, brothers, sisters, stepbrothers, stepsisters, aunts, uncles, nieces, nephews, and in-laws.

(e) No VE may administer or certify any examination by fraudulent means or for monetary or other consideration including reimbursement in any amount in excess of that permitted. Violation of this provision may result in the revocation of the grant of the VE's amateur station license and the suspension of the grant of the VE's amateur operator license.

(f) No examination that has been compromised shall be administered to any examinee. Neither the same telegraphy message nor the same question set may be re-administered to the same examinee.

(g) Passing a telegraphy receiving examination is adequate proof of an examinee's ability to both send and receive telegraphy. The administering VEs, however, may also include a sending segment in a telegraphy examination.

(h) Upon completion of each examination element, the administering VEs must immediately grade the examinee's answers. The administering VEs are responsible for determining the correctness of the examinee's answers.

(i) When the examinee is credited for all examination elements required for the operator license sought, 3 VEs must certify that the examinee is qualified for the license grant and that the VEs have complied with these administering VE requirements. The certifying VEs are jointly and individually accountable for the proper administration of each examination element reported. The certifying VEs may delegate to other qualified VEs their authority, but not their accountability, to administer individual elements of an examination.

(j) When the examinee does not score a passing grade on an examination element, the administering VEs must return the application document to the examinee and inform the examinee of the grade.

(k) The administering VEs must accommodate an examinee whose physical disabilities require a special examination procedure. The administering VEs may require a physician's certification indicating the nature of the disability before determining which, if any, special procedures must be used.

(l) The administering VEs must issue a CSCE to an examinee who scores a passsing grade on an examination element.

(m) After the administration of a successful examination for an amateur operator license, the administering VEs must submit the application document to the coordinating VEC according to the coordinating VEC's instructions.

§ 97.511 Examinee conduct.

Each examinee must comply with the instructions given by the administering VEs.

§ 97.513 VE session manager requirements.

(a) A VE session manager may be selected by the VE team for each examination session. The VE session manager must be accredited as a VE by the same VEC that coordinates the examination session. The VE session manager may serve concurrently as an administering VE.

(b) The VE session manager may carry on liaison between the VE team and the coordinating VEC.

(c) The VE session manager may organize activities at an examination session.

§§ 97.515-97.517 [Reserved]
§ 97.519 Coordinating examination sessions.

(a) A VEC must coordinate the efforts of VEs in preparing and administering examinations.

(b) At the completion of each examination session, the coordinating VEC must collect applicant information and test results from the administering VEs. The coordinating VEC must:

(1) Screen collected information;

(2) Resolve all discrepancies and verify that the VE's certifications are properly completed; and

(3) For qualified examinees, forward electronically all required data to the FCC. All data forwarded must be retained for at least 15 months and must be made available to the FCC upon request.

(c) Each VEC must make any examination records available to the FCC, upon request

(d) The FCC may:

(1) Administer any examination element itself;

(2) Readminister any examination element previously administered by VEs, either itself or under the supervision of a VEC or VEs designated by the FCC; or

(3) Cancel the operator/primary station license of any licensee who fails to appear for readministration of an examination when directed by the FCC, or who does not successfully complete any required element that is readministered. In an instancce of such cancellation, the person will be granted an operator/primary station license consistent with completed examination elements that have not been invalidated by not appearing for, or by failing, the examination upon readministration.

§ 97.521 VEC qualifications.

No organization may serve as a VEC unless it has entered into a written agreement with the FCC. The VEC must abide by the terms of the agreement. In order to be eligible to be a VEC, the entity must:

(a) Be an organization that exists for the purpose of furthering the amateur service;

(b) Be capable of serving as a VEC in at least the VEC region (see appendix 2) proposed;

(c) Agree to coordinate examinations for any class of amateur operator license;

(d) Agree to assure that, for any examination, every examinee qualified under these rules is registered without regard to race, sex, religion, national origin or membership (or lack thereof) in any amateur service organization;

§ 97.523 Question pools.

All VECs must cooperate in maintaining one question pool for each written examination element. Each question pool must contain at least 10 times the number of questions required for a single examination. Each question pool must be published and made available to the public prior to its use for making a question set. Each question on each VEC question pool must be prepared by a VE holding the required FCC-issued operator license. See §**97**.507(a) of this part.

§ 97.525 Accrediting VEs.

(a) No VEC may accredit a person as a VE if:

(1) The person does not meet minimum VE statutory qualifications or minimum qualifications as prescribed by this part;

(2) The FCC does not accept the voluntary and uncompensated services of the person;

(3) The VEC determines that the person is not competent to perform the VE functions; or

(4) The VEC determines that questions of the person's integrity or honesty could

compromise the examinations.

(b) Each VEC must seek a broad representation of amateur operators to be VEs. No VEC may discriminate in accrediting VEs on the basis of race, sex, religion or national origin; nor on the basis of membership (or lack thereof) in an amateur service organization; nor on the basis of the person accepting or declining to accept reimbursement.

§ 97.527 Reimbursement for expenses.

VEs and VECs may be reimbursed by examinees for out-of-pocket expenses incurred in preparing, processing, administering, or coordinating an examination for an amateur operator license.

Appendix 1 to Part 97—Places Where the Amateur Service is Regulated by the FCC

In ITU Region 2, the amateur service is regulated by the FCC within the territorial limits of the 50 United States, District of Columbia, Caribbean Insular areas [Commonwealth of Puerto Rico, United States Virgin Islands (50 islets and cays) and Navassa Island], and Johnston Island (Islets East, Johnston, North and Sand) and Midway Island (Islets Eastern and Sand) in the Pacific Insular areas.

In ITU Region 3, the amateur service is regulated by the FCC within the Pacific Insular territorial limits of American Samoa (seven islands), Baker Island, Commonwealth of Northern Mariana Islands, Guam Island, Howland Island, Jarvis Island, Kingman Reef, Palmyra Island (more than 50 islets) and Wake Island (Islets Peale, Wake and Wilkes).

Appendix 2 to Part 97—VEC Regions

1. 1. Connecticut, Maine, Massachusetts, New Hampshire, Rhode Island and Vermont.
2. 2. New Jersey and New York.
3. 3. Delaware, District of Columbia, Maryland and Pennsylvania.
4. 4. Alabama, Florida, Georgia, Kentucky, North Carolina, South Carolina, Tennessee and Virginia.
5. 5. Arkansas, Louisiana, Mississippi, New Mexico, Oklahoma and Texas.
6. 6. California.
7. 7. Arizona, Idaho, Montana, Nevada, Oregon, Utah, Washington and Wyoming.
8. 8. Michigan, Ohio and West Virginia.
9. 9. Illinois, Indiana and Wisconsin.
10. 10. Colorado, Iowa, Kansas, Minnesota, Missouri, Nebraska, North Dakota and South Dakota.
11. 11. Alaska.
12. 12. Caribbean Insular areas.

13. Hawaii and Pacific Insular areas.

DAMAGE NOTED

Torn area
Outside Back cover
2-9-'16

Made in the USA
Lexington, KY
02 September 2013